Anti-Corruption and its Discontents

T0199738

The fight against corruption is now a core part of development policy and practice. Some call these efforts a 'war on corruption'. What does this so-called 'war' mean for developing countries? And how do international perspectives on corruption relate to local and national concerns?

This book examines the relevance of anti-corruption discourse in Papua New Guinea (PNG), one of the most culturally rich and 'corrupt' countries on earth. Despite increased international, national and local efforts to address corruption over the past two decades, many fear that levels of corruption continue to rise largely unabated. Some believe that the mismatch between international, national and local assumptions regarding the nature of corruption and how it should be addressed is at the heart of the issue. International anti-corruption initiatives stress 'zero-tolerance' and try to strengthen formal state-based institutions. However, many people in PNG are more concerned about maintaining social relationships than following state laws and rules. This book critically examines the implications of the anti-corruption agenda and the collision of international, national and local perspectives. In doing so it provides a diagnostic on international assumptions about corruption and how it should be fought in developing countries, offering surprising and important lessons.

This book is essential reading for scholars and students of Development Studies, Geography, Political Studies and Economics, as well as practitioners and policy makers working in development.

Grant W. Walton is a Research Fellow for the Development Policy Centre at the Crawford School of Public Policy, Australian National University, Australia. Over the past decade Grant has conducted research in the Pacific, Asia and Africa. He has published in numerous academic journals and books and has authored major reports for donors and NGOs. He is the Deputy Director (International Development) for the Transnational Research Institute on Corruption, and a Research Fellow with the Developmental Leadership Program, University of Birmingham, UK.

Anti-Corruption and its Discontents

Local, National and International Perspectives on Corruption in Papua New Guinea

Grant W. Walton

Routledge
Taylor & Francis Group

LONDON AND NEW YORK

First published 2018 by Routledge

2 Park Square, Milton Park, Abingdon, Oxfordshire OX14 4RN
52 Vanderbilt Avenue, New York, NY 10017

Routledge is an imprint of the Taylor & Francis Group, an informa business

First issued in paperback 2019

British Library Cataloguing in Publication Data
A catalogue record for this book is available from the British Library

Library of Congress Cataloging in Publication Data
A catalog record for this book has been requested

ISBN: 978-1-138-69802-4 (hbk)
ISBN: 978-0-367-24522-1 (pbk)

Typeset in Times New Roman
by Taylor & Francis Books

For Anna, Marcel and the next little Walton

Contents

Illustrations

Acknowledgements

This book is about the differences between local, national and international perspectives on corruption in Papua New Guinea (PNG), a country located just to the north of Australia. The book draws on eight years of research, and multiple sources of data including large-scale surveys and more granular ethnographic work. It is argued that PNG is caught up in a broader international response to corruption, which has overlooked some local and national interpretations of corruption. The work is based – in part – on my PhD thesis (Walton, 2012) and there were many kind souls who have helped me to navigate the road to PhD completion and to the completion of this book. My primary supervisor Professor Jon Barnett often provided clarity and was patient and supportive throughout. My secondary supervisors Dr Monica Minnegal and Professor Leslie Holmes were also generous with their time and ideas. Peter Larmour and Ed Brown provided valuable feedback. Of course, all mistakes remain my own.

I am also thankful for the support of The University of Melbourne. The University provided a scholarship for three-and-a-half years of this process, which was topped up by funds secured by Professor Jon Barnett. The Department of Resource Management and Geography funded my fieldwork in PNG and attendance at a conference in Thailand.

My recent home, the Development Policy Centre at the Australian National University, has provided support for my continued research into issues on corruption and other challenges facing PNG. I am grateful to all of those I work with. In particular, thanks to Stephen Howes for allowing me the time to develop this book. I also thank the Developmental Leadership Program's Caryn Peiffer who has provided further insights into PNG perspectives on corruption through her adroit statistical analysis.

During my fieldwork in PNG for chapters 3, 4 and 5 of this book, fortuitously, many stars aligned. The University of Papua New Guinea was extraordinarily generous in providing accommodation for my wife and me. I thank Professor Betty Lovai for facilitating our visit. Professor David Kavanamur acted as a field supervisor in PNG and helped with letters of introduction and contacts. I was also fortunate to work for TI PNG during my time there. Thanks go out to the hard-working staff, including Emily Taule, Ivan Jemen, Marcus Pelto, Daniel George, Henry Yamo and Simon Jenkins.

Thanks to the Corruption Perception Advisory Group, which provided guidance on research into rural people's views on corruption; it was comprised of: Dr Sarah Dix (National Research Institute), Paul Barker (The Institute of National Affairs), Alois Francis (Consultative Implementation and Monitoring Council), Emily Taule (Transparency International Papua New Guinea), Dr Orovu Sepoe (University of Papua New Guinea), Fiona Hukula (National Research Institute) and Dr Alphonse Gelu (National Research Institute). I would also like to thank the numerous researchers and coordinators of the research.

Fellow PhD candidates Ahsan Rana, Jeffery Engels, Jenny Arnold, Michelle Hannah, Alexander Cullen, John Cox, Stuart Kent, Carmen Lindemann, Tanya Jakimow, Caroline Andrews, Laura Griffin, Shahar Hameiri and Suet Leng Khoo all provided much appreciated friendship, ideas, insight and comic relief, usually over a pot at the nearby Corkman pub. Friends outside of the university bubble – Fiona, Andrew M., Jess, Doug, Alex, Nickyy, Andrew, Imo, Sarah, Gus and the Millsteed crew – have also helped keep me sane and, crucially, made me realise that there is more to life than university.

My parents, sister and brother have been supportive, thank you. Thanks to mum for commenting on an early draft of the first and last chapters.

Anna Walton has provided endless support (including transcription and editing) and love throughout this long process. In recent years we have been joined by another little Walton, Marcel, who continues to bring us joy.

I would also like to thank the many people who participated in this research. Respondents from communities across the country as well as those associated with the PNG anti-corruption industry gave generously of their time and ideas. Thank you.

I am also grateful to the Journal of Development Studies for granting permission to use figures, tables, and some text from my published journal article (Walton, 2015), which features in chapter 3. And to Asia-Pacific Viewpoint for granting permission to reproduce text from a previous article (Walton 2013c).

Acronyms and abbreviations

ACW	Anti-Corruption Walkathon
AusAID	Australian Agency for International Development
CAID	Complaints and Administration Investigations Division (of the Ombudsman Commission of PNG)
CCAC	Community Coalition Against Corruption
The Coalition	The Non-Government Organisations and Civil Society Coalition
CPC	Constitutional Planning Committee
CPI	Corruption Perceptions Index
ECP	Enhanced Cooperation Program
EHMI	Eda Hanua Moresby Inc.
GCB	Global Corruption Barometer
GDP	Gross Domestic Product
ICAC	Independent Commission Against Corruption
IMF	International Monetary Fund
ITFS	Investigation Taskforce Sweep
LD	Leadership Division
LNG	liquefied natural gas
LPV	Limited Preferential Voting
MP	Member of Parliament
NASFUND	National Superannuation Fund
NCD YA	National Capital District Youths Association
NGO	Non-Governmental Organisation
NIO	National Intelligence Officer
NPF	National Provident Fund
OC PNG	Ombudsman Commission of Papua New Guinea
ODE	Office of Development Effectiveness
PM	Prime Minister
PNG	Papua New Guinea
PNG MGGO	Papua New Guinea Millennium Good Governance Organisation
PNG NAF	Papua New Guinea National Awareness Front
TA	Technical Assistance

TI	Transparency International
TI PNG	Transparency International Papua New Guinea

Papua New Guinean Currency Exchange

1 PNG Kina = 0.4 Australian Dollars; 0.32 USD; 0.3 Euro (exchange rate, January 2017).

1 Introduction

Corruption is a contested concept because individuals and communities frame it through very different worldviews, values and experiences. Activities that some individuals and communities believe constitute corruption (for example, giving gifts or employing relatives without due process) can be seen by others as culturally acceptable (Yang, 1989, 1994; Shore & Wright, 1997; Haller & Shore, 2005b). The problem is exacerbated through the constraints of language. The English term 'corruption' has a number of different meanings; it can refer to a type of transaction (a bribe, for example), not following correct procedure (for example, illegal graft), a type of moral decay (as in the phrase 'the corruption of youth') or a computer file that is no longer working properly ('a corrupted file'). In addition, in some languages there is no effective equivalent for the word corruption. These multiple meanings can make it difficult to discern exactly what people mean when they refer to corruption.

Yet the contested nature of corruption is rarely acknowledged in the expanding anti-corruption industry (a group of interconnected actors including donors, NGOs, think tanks and others, who are focused on addressing corruption around the world). For many working in the industry corruption is a universal concept and practice that must be addressed wherever it is found (Klitgaard, 1991; Pope, 2000a; Lambsdorff, 2007). This has led many to question the industry's legitimacy, with some arguing that the industry fails to acknowledge the relative and changing nature of corruption (Haller & Shore, 2005b; Sampson, 2005; Bukovansky, 2006; Sampson, 2010). But both advocates of universal interpretations of corruption, and their critics, still have little evidence with which to support their arguments. The literature on corruption has been growing, but there have been few attempts to empirically compare the ways anti-corruption organisations and citizens understand the concept in a meaningful way.

That is not to say that there have not been some attempts to do so. For example, through comparing responses to their 'Global Corruption Barometer' (GCB) and 'Corruption Perceptions Index' (CPI), which draw on public and elite views respectively, Transparency International (TI) argue that elite perceptions correlate with those of citizens in developing countries (Transparency International, 2009). However, both studies require respondents to interpret the word 'corruption' for themselves, and so offer no firm basis for comparing perspectives – because (as we shall see in chapter 2) definitions applied to this term vary widely. Thus, while scholars and practitioners acknowledge differences and similarities between elites' and citizens' perspectives on corruption, there are still gaps in our knowledge about these differences due to the limitations of the methodologies used.

Moreover, there have been few academic studies that have directly compared the views and practices of anti-corruption organisations. Most researchers focus on a single type of anti-corruption organisation when discussing their concerns about (or support for) the industry. For example, Bukovansky (2006) focuses on international donors, and Hindess (2005) examines TI, but neither compares different types of anti-corruption organisations. Given vast differences between local, national, and international anti-corruption organisations, drawing conclusions about the industry based on the study of one organisation, or type of organisation, can potentially be misleading.

The paucity of information about how different groups and individuals understand corruption has theoretical and practical implications. Theoretically it means that explanations of corruption, and of the work of the anti-corruption industry, suffer for a lack of evidence. Without empirical findings that compare anti-corruption and citizens' perspectives on corruption it is difficult to test theories about the legitimacy of the anti-corruption industry and the nature of 'corruption' itself. Practically, it means that anti-corruption practices may be ineffectual. For example, if anti-corruption organisations fail to account for the cultural, social and economic importance of transactions they label corrupt, it is possible that their policies will lead to local resistance through deliberate non-compliance.

The aim of this book is to compare international, national and local perspectives on corruption in Papua New Guinea (PNG) – a country where, like many developing countries, corruption is considered to be a significant problem and where international and national organisations struggle to address it. In this research international perspectives are gathered from two international anti-corruption organisations working in PNG, while national perspectives are garnered from two national anti-corruption organisations. Local perspectives are represented from qualitative and quantitative research involving urban and rural Papua New Guineans.

The war on corruption

Many within the anti-corruption industry see themselves at war. Anti-corruption heavyweights, the World Bank and the IMF, have called for a 'war against corruption' (World Bank, 2001). TI describes itself as 'leading the fight against corruption' (Transparency International Papua New Guinea, 2009a). In PNG – the site of this study – the media, supported by a range of civil society groups, declared a 'war on corruption' in January 2002 (Keaeke, 2003).

The war against corruption is ostensibly a war against poverty, crime and state instability. Andrew Vorkink, World Bank Country Director for Bulgaria and Romania, states that the 'war against corruption is a fight against poverty, a fight against the undermining of the rule of law and a fight against instability' (Vorkink, 2000: 1). Indeed the deleterious outcomes of corruption are well publicised. Based on surveys of enterprises, the World Bank estimates that bribery costs approximately US$1 trillion per year, worldwide (World Bank, 2010: 1). The Bank stresses that this is only a partial representation of the global costs of corruption, as it does not include the costs of acts such as embezzlement, fraud and nepotism. The Bank estimates the opportunity costs of corruption too, calculating that:

> countries that improve on control of corruption and rule of law can expect (on average), in the long run, a four-fold increase in incomes per capita ... such a country could expect, on average, a 75% reduction in child mortality.
>
> (World Bank, 2010: 1)

Given these statistics, the billions spent in the name of tackling corruption appear justified.

But the 'war on corruption' may do little to solve problems associated with corruption. Like the 'war on drugs' or the 'war on terror', the 'war on corruption' assumes a clear threat exists from an identifiable and intentional enemy. However, this is not the case. Indeed, fighting corruption is fraught with difficulties for four key reasons. First, those who are corrupt conduct their business in private and do not self-identify through membership of identifiable groups (such as armies). Second, definitions and legal renderings of corruption are contested and change throughout time (Scott, 1972). Third, understandings of the types of acts that are associated with corruption vary greatly between nations, communities and individuals (Gupta, 1995; Williams, 1999; Haller & Shore, 2005a). Finally, not all are convinced that corruption is, in itself, bad. Some even argue that acts considered corrupt are necessary for economic development (Nye, 1967; Huntington, 1989), state stability (Arriola, 2009) and the maintenance of cultural relationships (Yang, 1989, 1994). So, the enemy is contested, amorphous and changing all the time, and the sources and nature of the problem ('corruption') are themselves ambiguous; this raises the question of the suitability of the 'weapons' used in the 'war' against corruption.

Situating this book: geographies of corruption

Despite the growing international interest in corruption there has been little scholarship by geographers on the subject – notwithstanding a few notable exceptions (e.g. Perry, 1997, 2001; Brown & Cloke, 2004, 2005). This is concerning because corruption is shaped by issues that geographers are concerned with: politics, culture, environment and economics. Corruption also shapes natural and man-made environments. Despite the topic being under-studied in geography there are traces of scholarship that point to what 'geographies of corruption' might look like. Featuring geographers and others, this scholarship highlights how corruption and its narratives vary by place, scale and space.

One of the few geographers who have been concerned with corruption, Peter Perry, suggests that corruption shapes, and is shaped by, place. In his examination of political corruption in Australia, Perry (2001) argued that corruption has played a critical role in shaping politics, the economy and the landscape throughout the country's history. Other geographers have also noted the corrosive effects that corruption has had on the natural environment in developing countries (Laurance et al., 2011). While local conditions shape the extent and manifestations of corruption, they also impact on corruption narratives. Anthropologists and political scientists have noted that cultural, economic and social relations shape people's understandings about how corruption is interpreted. The fine line between a bribe and a gift often depends upon local context, rather than a set of universal laws and rules.

Corruption and its narratives also have spatial dimensions. Corruption is linked to transnational crime networks that cross country and regional boundaries. Perceptions about corruption similarly traverse space, as demonstrated by the Arab Spring where social media carried stories about small- and large-scale corruption. In Tunisia, the story of a young fruit-seller protesting against bribery and Wikileak cabals describing large-scale corruption by elites galvanised protesters. The Arab Spring highlighted the geographical aspects of anti-corruption protests: rallies mostly occurred in larger cities rather than rural areas. In PNG the divide between urban and rural locations is acute, with state services (including anti-corruption organisations) concentrated in urban areas. This book examines how the split between rural and urban locations impacts on perspectives on corruption in PNG.

Finally, scale – a key theme of this book – affects and is affected by corruption. Cannon (1966, cited in Perry, 1997) shows how the rail network – an invention that compressed space – of Victoria was shaped by corruption. In his recent book, anthropologist Akhil Gupta (2012) reminds us of the importance of scale for understanding and addressing corruption:

> Where modern bureaucracies are concerned, even when an unambiguous legal mechanism exists to determine corruption, if there is no widespread social agreement about which scale is to be used to judge correct ethical behaviour, the social judgement of corruption can often be contentious and fractured. Judgements about what kind of behaviour is corrupt do not simply diverge along class lines; even from the perspective of the poor no single viewpoint exists by which behaviour is judged corrupt.
>
> (81)

In other words, international and regional agreements that are drawn up inside the United Nations headquarters in New York are essentially different to the granular understandings of corruption in rural PNG, for example. Anti-corruption efforts also occur at different scales with anti-corruption actors working at international, regional, national and local levels.

Geography is crucial to understanding corruption, yet much of the corruption literature treats corruption as though it is immune to the vagaries of place, space and scale. This book, in its own small way, seeks to address this imbalance by examining how perspectives on corruption are constituted by these geographical concepts.

An introduction to corruption in PNG

The problem of corruption in PNG is said to be acute. External assessments of the problem – such as TI's CPI and the World Bank's World Governance Indicators – suggest that PNG is one of the most corrupt places on earth. For example, in 2016 PNG was ranked as the 136th most corrupt country (out of 176); on a scale of 0 for very corrupt and 100 for clean, PNG scored 28. Inside PNG there is also a great deal of consternation about perceived growing levels of corruption, with citizens taking to the streets, the media exposing corruption scandals and the anti-corruption industry highlighting the inadequacies of government responses to the problem. Listening to all of this discourse can make anyone feel overwhelmed and believe that the situation is hopeless; many believe that corruption will ultimately gobble up any riches that PNG has and send it spiralling into the status of a 'failed state'. But there are dim fragments of light that filter through the cracks of this ostensively crumbling state.

One illuminating light has been the (now defunct) anti-corruption agency, Taskforce Sweep. Set up by the Prime Minister, Peter O'Neill, in 2011 the taskforce was successful in recovering tens of millions of dollars' worth of stolen money and served arrest warrants upon a number of high-profile businesspeople and politicians. It even tried to help serve an arrest warrant on O'Neill himself, which resulted in him cutting off funding to the agency, leading to its ultimate demise at the end of 2014. In other words, Taskforce Sweep was a victim of its own success.

I have met with the head of the taskforce, Sam Koim, on a number of occasions and even interviewed him a couple of times for the Development Policy Centre's (my employer at the Australian National University) daily blog (http://devpolicy.org). My first interview with Koim was in a noisy and cramped corner of a hotel in Geelong, a satellite town of Melbourne, Australia. He had a plane to catch so the interview was short, and went by rather quickly. But what struck me about that conversation was Koim's concern about the very material concerns facing Papua New Guineans and how these concerns helped perpetuate corruption and support for it. Koim said that corruption was, in part, brought about because people see, read and hear about money in the midst of poverty. In a communal society, the accumulation of wealth caused 'jealousies' and ruptures that resulted in greedy behaviour, he argued. He was also concerned about the cultural underpinnings of corrupt activities. Without a social security system most people in PNG rely on the *wantok* system – literally 'one talk', a system of obligation between kin – to get by. Koim said that the pressures from *wantoks* means that many public servants look for ways to augment their salary. For Koim, while this system of obligation helped cause corruption, it could also help address it. He argued that any new anti-corruption organisation in PNG must reflect local, traditionally held, values. Koim believed that fighting corruption in PNG would not be successful without local people's values being reflected in anti-corruption activity. And that, ultimately, fighting corruption is about addressing socio-economic and cultural issues that affect people's lives.

For me, it was not surprising to hear this from Koim, even though he is a trained legal expert, who worked as the Principal Legal Officer at the Department of Justice and Attorney General (and thus one might suppose him more likely to be less sympathetic to cultural/material issues and take a legal perspective towards the issue). However, it was refreshing given the popularity of 'zero tolerance' approaches and a war on corruption that have been undertaken without much success in the country – and given the insistence by some practitioners and academics that the way citizens interpret corruption matters for less than their willingness to follow the rule of law. In the rush to address corruption in PNG, these cultural and structural issues are often overlooked. Instead, the causes and solutions to corruption are often seen to lie in PNG's formal state institutions.

My interview with Koim highlights the key question that faces policy makers and academics in trying to get to the bottom of the corruption complex in the country. That is, why, despite campaigns and increased resources to fight corruption, do citizens continue to support corrupt politicians, turn a blind eye to corrupt acts and show ambivalence to anti-corruption efforts? Is there a fundamental difference in the assumptions guiding anti-corruption efforts and the perspectives of citizens? If so, can these differences be bridged, or does fighting corruption in PNG – and places like it – require a fundamentally different approach?

In its own modest way, this book seeks to answer these questions by examining the way international, national and local actors reflect different perspectives on corruption. To do so it draws on extensive research conducted right across the country. It presents the views of both those engaged in anti-corruption efforts in PNG, and those who are most affected by corruption – that is, Papua New Guinean citizens – and shows how these perspectives sometimes clash and other times coalesce with international perspectives. It explains what the differences between local, national and international concerns about corruption mean for our understanding about corruption, the effectiveness of anti-corruption efforts and the broader war on corruption. It argues that understanding these perspectives is crucial for evaluating anti-corruption efforts and, ultimately, developing meaningful responses to corruption. By critically examining the collision of local, national and international perceptions about corruption in PNG, it provides a diagnostic on international assumptions about corruption and how it should be fought.

A rough guide to 'the land of the unexpected'

Locals and foreigners alike describe PNG as the 'land of the unexpected'. It's a title that, in part, derives from the country's struggle to graft modern institutions onto fragmented, ancient tribal cultures. Today's PNG – situated to the north of Australia and to the east of the Indonesian-controlled West Papua – was shaped by European powers in the late nineteenth century. The Dutch annexed the western part of the island of New Guinea, while the eastern part of New Guinea was divided into two: the Germans conquered the north while the British controlled the south. Australia conquered German territory during the First World War and continued to administer both halves of the country until 16 September 1975, the date of PNG's independence – when the southern Territory of Papua merged with the northern Trust Territory of New Guinea.

PNG is one of the most culturally diverse nations in the world. It boasts between 715 and 823 languages (Rannells & Matatier, 2005: 95), and around 10,000 autonomous political units (mostly numbering between 500 and 1,000 people) commonly described as tribes, which are further divided into clans (Ketan, 2007: 14). In addition, land has been and still is communally held. The vast majority of the land is communally owned, yet this communal ownership has been under threat from both international development agencies and corruption. In the 1990s the World Bank, IMF and the United Nations pressured PNG to privatise land ownings as they considered communal tenure a barrier to foreign investment (Gosarevski, Hughes, & Windybank, 2004b). While these efforts were resisted in PNG, communal land holdings came under attack from corrupt land grabs by foreign-owned logging companies under a notorious land development scheme called the Special Agricultural Business Licences. Now communal land ownership has likely dropped from the often-quoted figure of 97 per cent.

The German, British and Australian administrations brought centralised government, industrial technologies (including guns, aircraft and ships), Christianity, capitalism, Western medicine and education. At various times local people resisted these changes, and other times they acquiesced. The first recorded strike was held in the then northern capital, Rabaul, in 1929. It involved 3,000 labourers, including 200 of the 217 police stationed in Rabaul (Jinks, Biskup, & Nelson, 1973: 244). Strikers ended their protest peacefully, but received harsh punishment for their involvement – police were imprisoned and dismissed from the force, and some civilian leaders were sentenced to three years' imprisonment. While resistance to colonial rule was sporadic, it still played an important role in shaping the nation's history.

The colonial experience brought some benefits, which have been embraced by many, but the colonial legacy is in many ways poor. PNG is a Christian country, with a great respect for the work of the churches. Central government has reduced violence and mistrust between clans, although conflagrations still flare up, particularly in the central Highlands region. The Australian colonial administration was often cautious towards development, and at times it was paternalistic towards the local population. By the time of independence, education was for the male few and little infrastructure had been built. While the colonial administration fretted about the corrosive influence of outsiders on Papua New Guinean culture, it did little to develop the country. The legacy of these paternalistic colonial policies is evident today. There are no railways, no nationwide network of roads. On top of this, airflight is expensive and often dangerous, making trade and service delivery difficult and costly.

Despite these challenges, and the dire predictions of many at independence, PNG has managed to survive. While oftentimes challenged, the state has not failed; PNG continues to operate as a constitutional parliamentary democracy, with an elected Prime Minister as head of the government. Although this system of government has been grafted onto a complex socio-political environment, some claim it has served the nation well. With its unicameral parliament, PNG has been one of the few post-colonial states to maintain an unbroken record of democratic government (Reilly, 1999). In addition, PNG's judiciary has earned, and maintained, a reputation for independence. In 2005, the Supreme Court overruled a decision to allow Australian police to work in the country without being subject to local laws. The highest court in the land considered the actions of the then PNG government and a key foreign power as unconstitutional. Similarly, in 2016 the Supreme Court ruled that the country's detention centre for Australia-bound refugees was also illegal, throwing Australian refugee policy into great uncertainty weeks before the mid-year Australian election.

PNG has also managed to achieve modest economic growth, in fits and spurts, throughout its post-colonial history. In the early years of independence, policy reforms were introduced, which included a favourable renegotiation of the Bougainville Copper Mining agreement (which contributed substantially to the national economy). The incumbent government was re-elected in 1977 and there were smooth changes of government in 1980, 1982 and 1985 (May, 2004). However, by the mid-1980s the government's ability to deliver services noticeably declined. By the 1990s there were increasing problems with crime and lawlessness, a civil war on the mineral-rich land of Bougainville and clear signs of poor economic management. May (2004) suggests that it was during the 1990s that nepotism and corruption became noticeable. The nation's economy significantly deteriorated to the point where, in 2006, the United Nations' Committee for Development Policy unsuccessfully argued for PNG – along with Zimbabwe – to be downgraded to the status of 'least developed country' due to a low gross national income, economic stagnation and poor prospects for future growth (Committee for Development Policy, 2006: 28–29).

Since then PNG's economic performance has improved. The country's per-capita gross domestic product (GDP; at constant 2010 prices) increased from US$1,153 in 2002 to US$1,783 in 2014 – still well below the world average of US$10,242, and US$307 less than its nearest neighbour the Solomon Islands (World Bank, 2017). Economic growth has been higher than the global average since the 2007/08 global financial crisis. The country is set to build upon this economic fecundity, although PNG is never far from crisis. A major liquefied natural gas (LNG) project has recently been completed, which initially promised a direct capital investment of 36 billion kina (US$11.3 billion) in real terms over the life of the project (ACIL Tasman, 2009: v). It was also anticipated to double GDP, public consumption and foreign currency exports, and lead to a 45 per cent increase in aggregate employment (ACIL Tasman, 2009: viii). The project created great expectations that the country would reap the benefits of this enormous project. However, with a recent reduction of oil and gas prices, growth forecasts have been downgraded. PNG's GDP grew by 9.9 per cent in 2015 (Asian Development Bank, 2016), which represents a significant downgrading from an earlier estimate of 21 per cent from the Asian Development Bank (Asian Development Bank, 2015; Brennan, 2015). These growth figures mean that PNG has been one of the faster-growing economies in the world, yet it is a let-down from the great expectations associated with the LNG project.

Still, the country is becoming increasingly economically autonomous. This is reflected in its reduced reliance on foreign aid. At constant 2011 prices, between 1990 and 2012 official development assistance per capita more than halved – reducing from US$210 to US$93 (World Bank, 2014). This has been reflected in a declining reliance on Australian aid in particular. Indeed, at the time of PNG's independence, Australian aid represented 40 per cent of PNG's government budget; in 2015 it represented around only 8 per cent. This is despite the fact that Australia remains PNG's largest bilateral aid donor (Department of Foreign Affairs and Trade, 2015). PNG is committed to continuing to reduce its reliance on foreign aid, particularly from its largest donor, Australia (Abal & Smith, 2010), with some in both Australia and PNG announcing they want an exit strategy for Australian aid (Independent Review Team, 2010). Politicians in particular have made no secret that one of the key goals of economic expansion is to become economically independent from Australia. Ex-member of Parliament (MP) Bart Philemon has said that expanding economic growth means that, 'if we have problems with Australia, particularly, we [can] say, "well you're not the only one, we can go elsewhere"' (Fauziah, 2009). Like never before PNG is standing on its own two feet.

Yet the challenges facing this relatively young nation are enormous. According to the most recent analysis, rates of poverty did not change between 1996 and 2011: 40 per cent of the population lived under the poverty line. In 2015 the country scored 0.505 (out of 1), and was ranked 158th out of 188 countries in the United Nations' Human Development Index (HDI) (United Nations Development Program, 2015). Despite PNG's Millennium

Development Goals being customised to make them more achievable, neither the national or international goals were achieved by 2015.

To tackle these problems the government will need to ensure the effectiveness of every dollar it can get its hands on. But like many developing countries, corruption has become central to explanations of PNG's poor social progress. It has been considered a threat to business and investment (Hughes, 2003; Barker & Awili, 2007: 25), democracy (Standish, 2007; Allen & Hasnain, 2010), the sustainable use of natural resources (Barnett, 1989; The Asia-Pacific Action Group, 1990; Roberts, 2006), service delivery (Dorney, 2001) and law and order (Pitts, 2001, 2002). With numerous reported cases of corruption occurring in government, the private sector and NGOs (Okole & Kavanamur, 2003), it is self-evident to many that corruption is a problem in PNG and must be addressed.

In response there has been an increase in both formal and informal anti-corruption activities in the country. As described in chapter 4, following the trajectory of the international responses to corruption, the anti-corruption industry in PNG has expanded since 1997. Backed by significant funding, policies and institutions have been devised to address this issue. However, those working in the anti-corruption industry face a raft of factors – common in many developing countries – that problematise simplistic assumptions about the relationship between corruption and poor socio-economic indicators.

There may be very good reasons as to why Papua New Guineans engage in acts that outsiders label as 'corrupt'. With a paucity of services, no state-sponsored social security, few employment opportunities and little means for development, many communities on the periphery of 'Westernisation' in PNG wait in hope for development to arrive (Dwyer & Minnegal, 1998). It is for these reasons, Crocombe (2001) suggests, that while citizens may not approve of 'corruption', they may condone it, and rationalise it, if they benefit from it. Likewise, others have stressed cultural issues that propel Papua New Guineans into engaging in acts that some may view as corruption, as Kanekane (2007: 23) notes:

> Corruption in Papua New Guinea has been widely considered to be morally wrong and unacceptable. Yet, to some extent, it is an alien, Western-imposed notion... There is a dichotomy between what is per-ceived as corruption by outsiders and what is accepted by Papua New Guineans. Those who take the hard line that corruption is a black and white issue must realize that practices which are defined as corrupt may be culturally and socially accepted, or at least tolerated practices in Papua New Guinean societies.

These authors highlight the importance of understanding the structural con-straints on individuals and communities, and that doing so requires one to view the problem through subaltern eyes.

Those in the anti-corruption industry who fail to appreciate that corruption is, to some extent, contextually specific, leave themselves open to criticism. In PNG, many political leaders have argued that international anti-corruption organisations operating in the country are misguided (Rheeney, 2005; McPhedran, 2006). Indeed, in 2005 both the then Prime Minister, Sir Michael Somare, and the leader of the opposition (and, at the time of writing, the Prime Minister), O'Neill, publicly admonished the local chapter of Transparency International (TI PNG), suggesting that its statements about the level of corruption in PNG were unnecessarily tarnishing the nation's reputation (Rheeney, 2005). As I detail later in the book, the stateswoman, and at the time the only female serving in the PNG parliament, MP Dame Carol Kidu, has argued that many politicians may work outside of the rules and laws guiding their office because of their concern for their people, not for their own financial benefit. Kidu's plea is a popular rebuttal to anti-corruption organisations and outsiders who 'have been quick to label corruption as a major crime, without actually examining why it has increased in Papua New Guinea' (Kane-kane, 2007: 23).

Although debates about corruption in PNG have increased since the late 1990s, when I started examining this issue in the late 2000s little was known about how people in PNG conceptualised corruption. Surveys in cities such as Port Moresby provided insight into the degree to which some Papua New Guineans thought 'corruption' was a problem (PNG Justice Advisory Group, 2007), but could not explain how Papua New Guineans actually perceived what it meant. In particular, there was little information on the views of rural communities – where more than 85 per cent of the population live (Rannells & Matatier, 2005: 190). This is problematic as it is here where forms of exchange that prefer reciprocity over individual accumulation are strong. These informal and 'traditional' modes of exchange occur between individuals of the same clan, family or region. This is the *wantok* system described earlier, which is sometimes cited as a key reason why PNG suffers from corruption (Payani, 2000).

Similarly, little was known about anti-corruption organisations operating in PNG. There had been some insight into individual anti-corruption organisations (which, for the purposes of this analysis, include organisations that engage in a range of other activities besides anti-corruption – see chapter 5); for example, scholars had investigated TI PNG (Larmour, 2007, 2012), AusAID (Temby, 2007) and the Ombudsman Commission (de Jonge, 1998). However, there were no studies that compared anti-corruption organisations that operated in the country. Nor was there research that compared local, national and international perspectives on corruption. This book does both of these things by drawing on almost a decade of research in PNG: it compares the views of different anti-corruption organisations, and compares these to the views of PNG's citizens.

PNG offers an important case study into the challenges of addressing corruption because the problem is said to be acute, there is a concerted effort to address it, and yet, as in many developing countries, there is little evidence that it is being meaningfully addressed. This book will argue that this is because there is a misalignment between the ways anti-corruption organisations and locals understand and respond to 'corruption'. PNG is marked by a set of 'traditional' social and economic relations that are still important. Traditional values of communalism (manifest in the extremely high rate of communal land, the pervasiveness of the *wantok* system, and the high number of people living in remote rural locations that are still structured along traditional lines) lie in marked contrast to the values of the Western world from which the international anti-corruption industry has emerged. Thus, PNG is an ideal case study to highlight the differences between the perspectives of the international anti-corruption industry and 'local' people and national organisations. In this sense, PNG is a 'typical' case study, in that it represents an environment similar to those that anti-corruption organisations face in most developing countries. So, investigating the anti-corruption industry in PNG not only adds to our understandings of the industry in that country, but also adds to knowledge about the impact of global anti-corruption efforts around the world.

Reflexivity of an insider/outsider

My interest in PNG began in 2003, when I visited the country for a short trip to intern for an NGO run by the strong-willed and inspirational Matilda Koma. I returned again in 2004 to conduct research on the impacts of the Tolukuma Gold Mine, located 100 kilometres north of the capital Port Moresby. While these visits were short they got me hooked, and, longing to return, I thus jumped at the chance to undertake fieldwork for a PhD in the 'land of the unexpected'.

I got interested in corruption in an altogether different part of the world: Afghanistan. It was here, while I was working for the NGO ActionAid, where I was exposed to debates about corruption and development. Conversations around this topic were often fraught. On the one hand many were concerned about government corruption and wanted it addressed at any cost. On the other, were concerns about what crackdowns on corruption might mean for state-building and the complex relations between tribal communities and their leaders. My interest was further piqued by an article by the academic Elise Huffer (2005), who clearly outlined the potential mismatch between donor, government and citizens' perceptions of corruption in the Pacific.

Moving back and forth between the world of development and academia has marked my subsequent research for this book. In PNG I worked with the local chapter of TI for a short time (in 2008 and sporadically in 2009 and 2012) on a survey into citizens' perceptions of corruption. At the same time, the agency has been one of the case studies featured in this book. This means I am both an insider to the anti-corruption industry and, through my academic position and the fact I do not have an ongoing role in the industry, an outsider. This straddling of the world of development and academia requires a degree of reflexivity – a sensitivity to not being sucked in by the narratives of powerful insiders, while not giving over to a type of nihilistic cynicism which denies the problem that corruption poses for developing countries such as PNG. This straddling has the advantage of seeing the mechanics of the anti-corruption industry from the inside, and being privy to the constraints and pressures that frame narratives of corruption. I have attempted to provide as much of an objective account of the evidence as possible in appraising the organisations within the anti-corruption industry and those outside of it.

Book structure

In order to compare international and local understandings about and responses to corruption it is vital that we can differentiate between perspectives. The next chapter reviews the scholarly literature and develops a typology of corruption to guide analysis in the following chapters. This typology classifies views on corruption: the mainstream and alternative views. These views cover five perspectives on corruption, with the mainstream view including legal, public office and economic perspectives and the alternative view covering moral and critical perspectives. In short, the mainstream view focuses on corruption and the state and the alternative view looks beyond the state to the structural issues (such as culture, power relations and poverty) that shape ideas about corruption. These perspectives are differentiated by the way they define corruption, ascribe causation, and seek solutions to corruption. The book illustrates the usefulness of this typology by applying it to narratives about corruption in PNG.

Chapter 3 examines the evidence on corruption in PNG and delineates between perceptions about corruption at the local level, and concerns about corruption at the national and international scales. It draws on findings from a large-scale survey, focus groups and an analysis of national and international narratives on corruption. It highlights the different scales through which corruption is manifest, caused and interpreted within and beyond the nation-state of PNG. At the local level the alternative view is critical for understanding marginalised concerns about corruption. Given the weakness of the PNG state, responses to corruption are often – though by no means always – determined by cultural considerations and economic marginalisation, which reflect the moral perspective. The chapter shows how local dynamics can shape people's willingness to respond to corruption. At the national scale the tensions between the mainstream and alternative views on corruption are mirrored in political scandals and policy. At the same time economic globalisation has increased the chances for corruption between private sector and public sector elites. The chapter suggests that addressing corruption in PNG will require anti-corruption actors to understand and respond to both the mainstream and alternative views on corruption at a variety of scales.

Chapters 4 and 5 turn attention to responses to corruption in PNG. These chapters draw heavily on fieldwork conducted in PNG between 2008 and 2009, and the supporting material (such as grey materials from these organisations) presented is mostly from – or leading up to – this point in time. The way Papua New Guinean organisations and institutions have responded to the threat of corruption is explored in chapter 4. This chapter provides a history of concern about corruption in PNG. The nation's founding leaders incorporated concerns about culture and the corruption of Western organisations in their pre-independence report. These concerns were, in part, reflected in the nation's constitution which calls for a form of development that reflects 'Papua New Guinean Ways', an approach that stresses the importance of drawing upon Western, Christian and cultural institutions. Drawing on interviews, the chapter then compares two anti-corruption organisations' views on corruption. It shows that the vision of the country's founders has been poorly translated by the nation's constitutionally mandated Ombudsman Commission (the OC PNG), which has failed to provide meaningful sanction over the illegal and unethical conduct of state officials. Reflecting the alternative view, those within the OC PNG longed for more culturally aware responses to corruption that reflected Papua New Guinean Ways. The chapter then examines a second anti-corruption organisation called the Coalition. The Coalition were an amalgamation of grassroots NGOs, who regularly protested against corruption. Reflecting a type of 'negative nationalism' (Robbins, 1998), they considered PNG's multiple cultures as a weakness in the fight against corruption. In line with the mainstream view, they sometimes called for Western companies and Western advisors to provide solutions to PNG's developmental challenges. Thus, demonstrations of the alternative view were found where they might be least expected – among the middle-class of the OC PNG, rather than within the ranks of local anti-corruption NGOs.

Chapter 5 examines the way in which international organisations have reflected the mainstream and alternative views on corruption. International anti-corruption organisations started to reshape responses to corruption within PNG in 1997, shortly after concern about corruption became a central issue for development agencies. TI PNG's approach to addressing corruption incorporated Papua New Guinean voices through its chapter system, yet official discourse on corruption masked the concerns of stakeholders who were at times critical of the mainstream view. The Australian aid program provided a more technical response to corruption, through focusing its efforts on strengthening Western institutions. Ultimately, both organisations reflected the mainstream view, by focusing on instituting a legal-rational Weberian state, where PNG's state institutions are, by and large, free from cultural pressures, politics and private interests.

The ways the four anti-corruption agencies responded to *political* corruption are the focus of chapter 6. By drawing on a range of evidence, which includes events up to the time of finalising this book (2016), this chapter shows that the Australian aid programme struggled to meaningfully respond to political corruption, particularly in addressing the transnational transfer of funds from Australia to PNG. TI PNG and the OC PNG were at times cautious, but provided some critique of and response to political corruption. However, it was the Coalition who provided a more direct and confrontational response to political corruption in PNG. As AusAID, TI PNG and the OC PNG were cautious in their approach to addressing political corruption, the Coalition were able to speak truth to power. However, the group's political entanglements and some of their solutions to corruption call for a sober assessment of the potential of their efforts to meaningfully address corruption.

The final chapter first compares the perspectives of national and international anti-corruption organisations and citizens' (local) concerns about corruption. It demonstrates that the assumptions framing responses to corruption in PNG are inadequate because they fail to account for marginalised perspectives on corruption, which are shaped by political, economic, social and cultural factors. The granular world of individuals and communities is often lost in the push to provide national and international responses to corruption. While national organisations are, in some cases, better adapted to meet these challenges, they do not always seek to empower their fellow citizens. It is argued that listening to what ordinary citizens say about corruption can lead to more meaningful solutions. In particular, structural constraints – culture, poverty and inequality – need to be acknowledged by anti-corruption agencies and others. This will also mean more meaningfully responding to the political context that frames corruption and responses to it. The chapter provides some suggestions as to how these concerns might be addressed within PNG.

2 Alternative and mainstream views on corruption in the land of the unexpected

Introduction

In PNG, as with many developing countries, corruption talk can be baffling for the uninitiated. In popular discourse the term can be used to describe a moral failing, illegal activity or institutional decay. To help untangle the variety of ways that Papua New Guineans talk about corruption, this chapter introduces a typology of corruption that comprises two different views – the mainstream and alternative – and five perspectives – legal, public office, economic, moral and critical – on corruption. The chapter highlights the way written narratives about corruption in PNG reflect these perspectives. While much has been written about corruption in PNG, few authors endeavour to say what they mean when they employ the term. Moreover, there have been few attempts to categorise narratives of corruption in PNG. This is problematic, for without clarity about how corruption is deployed, academics, policy makers, development workers and citizens end up talking at cross-purposes about the nature of corruption, and thus how it should be addressed. Peter Larmour highlights this problem by noting that the diversity of academic, official and popular conceptions of corruption in the Pacific region 'suggests there is great room for misunderstanding, irritation, and poor targeting of anti-corruption campaigns' (Larmour, 2006: 17).

Elise Huffer (2005) has provided a useful categorisation of definitions of corruption across the *Pacific* – as opposed to PNG per se – by drawing on the work of Robert Williams (1999). In doing so she argues that donors understand corruption through an economic lens, which has put them at odds with the moral interpretations of local communities. As a result, she calls for local communities to lead research about communal ethics across the Pacific to help rebalance the 'Western'-centric views of anti-corruption organisations. Using a similar, yet significantly expanded framework to Williams (1999) and Huffer (2005), [1] this chapter imposes some order on the narratives of corruption in PNG. In doing so, it finds that debates about corruption globally and in PNG are mostly framed by the 'mainstream' Western view of corruption.

First this chapter develops a theoretical framework for understanding the different views of corruption and anti-corruption in the international literature; it categorises the literature into mainstream and alternative views. It shows how these views are comprised of five perspectives that are differentiated by the way corruption is defined, causation ascribed and solutions proffered. This framework for understanding corruption reverberates throughout the rest of the book: it is drawn upon to assess the ways in which anti-corruption actors and citizens understand and respond to corruption. The chapter then shows how these views have been presented in debates about corruption in PNG by presenting a review of the literature in PNG, highlighting how these perspectives are deployed by different actors. The chapter concludes by suggesting that while the alternative view may be unpopular, it could provide important insights into the effects of anti-corruption work in the country, and could show how corruption may be better addressed.

Building a framework to understand corruption

There are four prominent definitions that underlie the bulk of the literature on corruption: moral, public office, economic and legal (Williams, 1999). These definitions have influenced the way academics and anti-corruption organisations have understood and responded to corruption. In addition, over the past ten years a growing 'critical school' has sought to reconceptualise understandings of corruption and the anti-corruption industry (Michael, 2004). These four definitions and emerging approach form the basis of the typology developed for this book. Each of these perspectives is differentiated by the way they interpret how corruption is defined, causes are ascribed, and solutions prescribed. The importance of each of these components is explored below.

Definitions

Definitions of corruption are an important starting point from which to understand this concept as they reveal the values and norms that are privileged in narratives about it. Many definitions of corruption have attempted to remain 'value-free' as they 'have typically been used by social scientists, especially economists, who do not want to prejudge the moral value of certain kinds of behaviour (bribery, embezzlement, etc.)' (Qizilbash, 2001: 267). However, as a number of academics have noted (Qizilbash, 2001; Hindess, 2005), no definition is truly neutral.

Indeed, popular definitions of corruption can exclude morally dubious behaviour that some consider corrupt. Many theorists, for example, define corruption as 'the abuse of entrusted power for private gain' (known as the 'entrusted power definition' in the remainder of this book). Political leaders have used this definition to avoid blame for their illicit actions, as Haller and Shore (2005b: 6) explain:

> When former German Chancellor Helmut Kohl disclosed in 2000 that he had used secret funds to finance his Christian Democratic (CDU) party

in the former GDR [German Democratic Republic] he insisted (drawing on the strategically useful IMF definition) that this was not 'corruption' as he had not made any private or personal gain.

Chancellor Kohl's response indicates that definitions can be drawn upon to justify wrongdoing by the powerful. For critics, definitions of corruption that include 'private gain' fail to acknowledge a great number of activities that may involve an abuse of entrusted power (Haller & Shore, 2005b; Hindess, 2005). This highlights the inescapably value-laden nature of the term 'corruption', and the importance of critically examining the limitations of its definitions.

While definitions set out the conceptual parameters of 'corruption', there has been a tendency in the anti-corruption literature to downplay the importance of discussing the limitations of its definitions (Klitgaard, 1991; Lambsdorff, 2007). For example, Johann Graf Lambsdorff (2007), the founder of Transparency International's (TI) CPI, suggests that those who spend time debating definitions of corruption waste energy that should be used to understand its causes and how to address it. Lambsdorff (2007) implicitly suggests that definitions of corruption are somehow disconnected from their causes and solutions. But, as definitions of corruption frame what may or may not be included in a corrupt activity (which determines what should be addressed), defining corruption must be a key concern for those who seek to address it. Moreover, while many scholars discuss possible causes of, and solutions to, corruption, the ways in which these are determined are far more subjective than many admit, as explained below.

Causes of corruption

Identifying the causes of corruption has become an integral part of scholarship on the subject. Corruption is caused by a range of diverse factors, including poverty (Gupta, Davoodi, & Alonso-Terme, 2002), weak government institutions (Payani, 2000), nationalism (de las Casas, 2008) and resource extraction (Marshall, 2001). But 'it is often difficult to assess whether corruption causes other variables or is itself the consequence of certain characteristics' (Andvig & Fjeldstad, 2001: 61). Thus, the asserted causes of corruption often reflect a priori assumptions that authors make about whether corruption causes certain phenomena (e.g. poverty) or is caused by them.

This has caused some academics to question the motivations of those who prefer some causes over others. For example, while there is some evidence that anti-corruption organisations can contribute to corruption (Anechiarico & Jacobs, 1996; Hanlon, 2004; de Maria, 2005), those associated with the industry tend to focus on the ways developing countries' governments cause corruption. Some scholars suggest that those associated with the industry also give scant regard to the private sector's involvement in corruption, thus supporting the neo-liberal agenda of anti-corruption organisations (de Maria, 2005; Brown & Cloke, 2011; Rocha, Brown, & Cloke, 2011). So, determining causation is contested, but is an important part of the literature and thus an important component of understanding perspectives on corruption.

Solutions to corruption

As discussed in chapter 5, writings on anti-corruption have increased exponentially over the past ten years (Michael & Bowser, 2009). There are many solutions to corruption proposed by anti-corruption organisations and academics. Solutions have included increasing women's participation in leadership positions (as it is assumed women are less corrupt than men) (Goetz, 2007), increasing the wherewithal of communities to monitor government- and private sector-managed projects (Richards, 2006), building a sense of nationalism in citizens (de las Casas, 2008), public education programmes (McCusker, 2006: 15) and strengthening state institutions responsible for accountability (i.e. the ombudsman, the judiciary, the executive) (Pope, 2000a). Again, like causes of corruption, solutions are contested and subject to a priori reasoning. For example, Krastev (2004) notes that solutions offered by academics are often determined by their disciplinary background. Others have argued that anti-corruption organisations proffer solutions that fit in with their own agenda. Chang (2008) argues that corruption is used as an excuse by developed countries to reduce their aid commitments, which has, in turn, exacerbated poverty in many countries.

Moreover, there are few examples of successful anti-corruption initiatives. Andvig and Fjeldstad (2001) state that Hong Kong and Singapore, two countries with authoritarian governments that were able to implement anti-corruption programmes with little opposition (and thus are rare examples), are the only 'clear-cut successes' of modern countries significantly alleviating corruption. After reviewing the literature, Hanna and colleagues conclude that implementing evidence-based anti-corruption reform is difficult because 'the body of micro-level empirical studies on anti-corruption interventions is extremely small at this point' (Hanna et al., 2011: 6). Thus – while research on corruption continues – there is insufficient evidence to definitively show which anti-corruption efforts effectively work, contributing to the contestation around how to address it. So, causes of, and solutions to, corruption are critical to perspectives on corruption; but both are contested, and can reveal as much about their proponents' values and goals as they do about the nature of corruption.

'Mainstream' and 'alternative' views on corruption

Theories about corruption abound, but a schism is growing between 'mainstream' and 'alternative' views on corruption. This section describes the way these views have been discussed in the international literature. In particular, it presents the different perspectives associated with each view, and describes how these perspectives define corruption, ascribe causation and prescribe solutions.

The framework employed here is influenced by the work of Williams (1999), Michael (2004) and Brown and Cloke (2011). Williams (1999) identified four *definitions* of corruption – legal, public office, economic and moral. Williams' (1999) discussion of these definitions is rather brief: he does not specify how these definitions frame the way causes are ascribed and solutions prescribed to corruption. So Williams' four definitions provide a basis for fleshing out broader 'perspectives' of corruption that include definitions of, causes of and solutions to corruption. Michael (2004) and Brown and Cloke (2011) have identified an important 'critical school' of scholarship. Their insights have contributed to the construction of the 'alternative view'.

Table 2.1 summarises the taxonomy developed for this chapter; it shows how the mainstream and alternative views, and their associated perspectives, define, ascribe causation, and prescribe solutions to corruption. It also identifies the scholars associated with these views. A description of each of these perspectives is provided next.

The 'mainstream' view

The mainstream view on corruption frames discussions about corruption in relation to the state. In this view corruption is understood in reference to state laws, rules and institutions, and therefore those who reflect this view particularly focus on the corruptibility of state officials. There are three key perspectives on corruption that align to this mainstream view. First, there is the *legal perspective on corruption*, which defines corruption as 'the violation of a rule' as determined by legal processes (Williams, 1999: 504). Corruption is, in this view, an act that breaches obligations set out in legal codes.

The legal perspective looks to inadequacies of the legal system to identify the causes of corruption. For example, Levin and Satarov argue that undeveloped legislation, legal loopholes, ill-defined legislative procedures and inadequate enforcement all contributed to the rise of corruption in Russia during the 1990s. Weak legal systems can result in a perpetuation of corruption, as Jain (2001: 72) illustrates:

> Once corrupted, the elite will attempt to reduce the effectiveness of the legal and juridical systems through manipulation of resource allocation and appointments to key positions. Reduced resources will make it difficult for the legal system to combat corruption, thus allowing corruption to spread even more.

In turn, the legal perspective focuses on addressing corruption through the passage and tightening of laws and regulations. Practical approaches based on this perspective include international measures such as the Organisation for Economic Cooperation and Development's (OECD) 1997 *Convention on Combating Bribery of Foreign Public Officials in International Business Transactions* and the United Nations' 2003 *Convention Against Corruption*.

Table 2.1 A summary of the mainstream and alternative views on corruption

View	Definition	Causes	Solutions	Scholars
Mainstream				
Legal	Corruption is the violation of the rule of law, based upon prima facie evidence.	Inadequacies of the legal system.	Reframing laws and ensuring better law enforcement.	Leff (1964); Scott (1972); Levin and Satarov (2000); Posadas (2000); Jain (2001).
Public office	The abuse of public office for private gain.	Ill-defined public and private spheres.	Public officials are supported (or coalesced) to adhere to their public roles.	Bayley (1966); Nye (1967); Treisman (2000); Van Rijckeghem and Weder (2001).
Economic	Corruption is rent seeking – returns in excess of a resource owner's opportunity cost that divert productive resources – primarily by state officials	Monopolistic control of resources by the state	Reduction of the size and power of the state through liberalisation, privatisation and deregulation	Klitgaard (1991); Rose-Ackerman (1999); Ali and Isse (2003); Goel and Nelson (2005)
Alternative				
Moral	Corruption means to 'pervert, degrade, ruin and debase' in line with community values.	Undermining 'traditional' and 'communal' values.	Solutions that emerge from the 'community'.	Yang (1989); Shore and Wright (1997); de Sardan (1999); Haller and Shore (2005a); Huffer (2005).
Critical	Definitions of corruption reflect the power of those who evoke it	Anti-corruption organisations cause corruption and other unintended consequences	Grassroots groups not associated with the anti-corruption industry – to re-politicise anti-corruption	Brown and Cloke (2004, 2011); Hindess (2005); Harrison (2006)

The success of these measures depends upon their ratification and effective implementation by nation-states, as it is at this level that legal instruments become crucial in addressing corruption (Posadas, 2000). Thus, from the legal perspective, addressing corruption requires both the implementation and enforcement of appropriate legislation.

Academics have long looked to the law to identify corrupt behaviour (Scott, 1972). For example, Nathaniel Leff, a scholar often associated with this perspective, regarded corruption as extra-legal activity. Many academics have taken Leff's lead, with Williams suggesting that modern social science has favoured an 'essentially legalistic understanding of the term' (Williams, 1999: 504). As a result, much 'debate about corruption is shaped by thinking in binary oppositions: corruption and the law are seen as each other's opposite' (Nuijten & Anders, 2007: 11).

The second perspective associated with the mainstream view is the *public office perspective on corruption*. Under this perspective corruption is defined as 'the abuse of public office for private gain'. This definition has been popular with academics and policy makers (World Bank, 1997, 2007; Treisman, 2000; Bukovansky, 2006; McCusker, 2006; Transparency International, 2015). Defined in this way, corruption has come to be connected to the state as, whilst 'public office' is a contested concept (Ruud, 2000), most literature on corruption has equated the term to government office (Hutchinson, 2005). In turn, this definition focuses upon the ways those who work for the state execute – or fail to execute – their duties (Scott, 1972). Nye suggests that corruption involves public officials, and that it can include behaviour such as:

> bribery (use of reward to pervert the judgement of a person in a position of trust); nepotism (bestowal of patronage by reason of ascriptive rela-tionship other than merit); and misappropriation (illegal appropriation of public resources for private-regarding uses).
>
> (Nye, 1967: 419)

In short, this definition suggests that corruption involves cheating the state (whether monetarily or otherwise) by its officials (although others, such as businesspeople, may also be involved).

Some have sought to widen the scope of this definition given the increasing importance the private sector plays in governance in the modern era. In 2000 TI, a bellwether for trends in the anti-corruption industry, changed the way the organisation defined corruption to 'the abuse of entrusted power for private gain' (Transparency International, n.d.-a). This change was designed to broaden understandings of corruption from acts that include the involvement of one or more state officials, to include improper business-to-business activity within the private sector (Holmes, 2006b), and corruption between members of civil society. Holmes (2006a) argues that broadening the concept of corruption was a response to the spread of neo-liberal policies since the 1970s, which was particularly accelerated in the 1980s; these policies have since helped blur the lines between government and private sector service delivery. In turn, this is a broad definition that moves away from corruption having to involve one or more state officials. Both of these definitions do, however, suggest that corruption involves those in positions of power – whether in the state or outside of it.

A number of anti-corruption organisations have employed this definition (Asian Development Bank, 2004; AusAID, 2007c). It is reflected in TI's internationally influential 'Bribe Payer's Index' (BPI) (Hardoon & Heinrich, 2011) and their Global Corruption Barometer, GCB, (Hardoon & Heinrich, 2013) – two annual reports designed to shame countries into improving their accountability and transparency. However, as Holmes (2006a) notes, the CPI, TI's most influential report, is still based upon the public office definition of corruption – i.e. corruption defined as 'the abuse of public office for private gain'.

While the 'abuse of entrusted power' definition has increasingly found its way into the policies and research instruments of anti-corruption organisations, there is still scepticism about the extent to which this shift in rhetoric has transformed into real efforts to mitigate private sector corruption (Brown & Cloke, 2005, 2006, 2011; de Maria, 2005; Holmes, 2006b). So, while this new definition is increasingly referred to in the anti-corruption industry, many still believe anti-corruption organisations are primarily (if not solely) concerned with corruption involving state officials (and thus reflect the 'public office' definition). This critical literature is further discussed later in the chapter.

From this perspective, causes of corruption are linked to a range of issues that give rise to public officials transgressing their official roles. Thomas and Meagher (2004) provide a useful categorisation of the types of causes considered by the public office perspective: they group causes of corruption into structural and individual kinds. Structural causes relate to the strength and capacity of the state. Thomas and Meagher (2004) posit that there can be no distinctive public office if the government is weak, there is no agreed public for the government to serve, and there is no clear separation between the public and the private spheres. Not having a clearly defined public office means accusations of corruption become inevitable (although problematic), as the definition of corruption from the public office perspective assumes there is an identifiable public office, and that it is segregated from the private sector and civil society.

The second categorisation identified by Thomas and Meagher (2004) concerns individual causes, which are concerned with the incentives that drive corruption. For example, Van Rijckeghem and Weder (2001) argue that low salaries force public servants to engage in corruption, while high salaries are a premium that is lost if they are caught and fired. Others suggest that individuals are motivated to engage in corruption if they are inadequately supervised or face insufficient sanctions (Polinksy & Shavell, 2001). Those advancing the public office perspective argue that without the right mix of (dis)incentives, corruption, particularly among public servants, will result.

Following on from the causes of corruption, solutions from the public office perspective are directed towards strengthening the state, augmenting state–society relations and changing the incentives for corrupt behaviour. The state may be strengthened through economic reform, oversight of public expenditure and financial management, support for legal and judicial systems, civil service reform and public oversight mechanisms (Michael, 2004). State–society reform may include the formation of anti-corruption coalitions between civil society groups and public officials. Reforms aimed at shaping individual behaviour can include internal monitoring and sanctions (through internal audit committees, reporting requirements and disciplinary codes, for example) and a broader set of positive incentives for correct behaviour (for example, increasing wages and rewarding merit-based behaviour) (Hamilton-Hart, 2001).

This approach is often guided by principal-agent theory, which focuses on the incentives for individuals to condone or resist corrupt behaviour. The principal-agent theory of corruption is based around two key actors: the principal (variously depicted as government ministers, agencies, or voters) and the agent (groups or individuals which the principal ostensibly monitors) (Ugur & Dasgupta, 2011; Sööt & Rootalu, 2012). Under this theory corruption occurs when information and preference asymmetry between principals and agents provides incentive for agents to engage in corruption. In other words, corruption occurs when principals are unable to adequately monitor agents and when the goals of the two are not aligned. The targets of this approach are most often state officials and politicians (who play the role of the agent), which plays into the concerns of the public office perspective that reforms should be targeted at strengthening systems of government and coercing those working within them to do the right thing.

The public office perspective came to the fore in the 1960s and dominated the literature until the 1980s (Williams, 1999). It is linked to scholars such as Bayley (1966), Nye (1967) and Myrdal (1968), as well as Treisman (2000) and Van Rijckeghem and Weder (2001) who have examined the effects of corruption on the public sector in a variety of cultural situations. While this perspective is still persuasive, since the 1990s, it has played second fiddle to the economic perspective, the third perspective associated with the mainstream view.

The economic perspective is associated with influential economists such as Rose-Ackerman (1999) and Klitgaard (1991) who view 'government through an economic prism' (Williams 1999: 506). For these authors, economics is the lens through which corruption can be understood and offers the best tools for addressing it. This perspective is centrally concerned about the productive use of economic resources. For example, Rose-Ackerman (1999: 2) states that 'endemic corruption suggests a persuasive failure to tap self-interest for productive purposes'. From this perspective corruption is often referred to as 'rent-seeking', which is defined as 'returns in excess of a resource owner's opportunity cost' (Williams 1999: 507). While this is a broad definition, the economic perspective is centrally concerned with the potential for government to monopolise resources and demand excessive rents.

This perspective suggests that the size of a government and its level of interference in an economy are key causes of corruption. There is some evidence to support this theory. For example, Ali and Isse (2003: 461), after reviewing studies on corruption in least developed countries, find corruption is 'significantly correlated to' the size of government; their model suggests that a 10 per cent increase in the size of government – measured as the government's expenditure share of GDP – leads to a 2 per cent increase in the perceived level of corruption. Similarly, Goel and Nelson (2005) examine the relationship between economic freedom (the degree of government intervention in the economy) and corruption. They argue that the less governments intervene in the economy, the lower the level of perceived corruption.

In turn, reducing the monopolistic tendencies of the state and increasing competition in the private sector are key remedies advocated by the economic perspective. To achieve this, Ohnesorge (1999) suggests that governments prioritise economic deregulation, liberalisation and privatisation. These policy responses were drawn upon by international agencies operating in Africa during the 1990s (Szeftel, 1998). Theobald (1990: 158), a supporter of such remedies, suggests that in Africa 'it may be that privatisation offers the only viable prospect of curtailing corruption'.

In summary, the mainstream view draws on the legal, public office and economic perspectives, all of which are concerned with corruption involving the state and state officials. This view has become particularly important for the way development organisations have responded to corruption. Responses to corruption under the this mainstream view have become quite technical in nature, with responses focusing on monitoring state officials and politicians, and right-sizing the state. This view is contested by those reflecting the alternative view, a view that takes a more expansive approach to understanding and responding to corruption.

The 'alternative' view

The alternative view of corruption acknowledges that understandings of corruption depend upon community norms and relationships of power. In turn, it focuses on local responses to address corruption. There are two perspectives associated with this view – the moral perspective and the critical perspective. The moral perspective brings a bottom-up approach to understanding and responding to corruption, being grounded in anthropological enquiry. Harrison (2006) suggests that the moral perspective has its roots in the earlier work on exchange and solidarity networks by authors such as Barth (1978), Mauss (1990) and Sahlins (1972) who – while not directly addressing corruption – explained the relational expectations that tied individuals to transactions with others. Given this, the moral perspective has been particularly popular among anthropologists who have written about corruption over the past two decades (Shore & Wright, 1997; de Sardan, 1999; Haller & Shore, 2005a).

From this perspective, corruption is sometimes defined as a form of individual or institutional decay. This is reflected by the online Oxford Dictionary, which defines corruption as a 'process of decay' or 'putrefaction' (Oxford Dictionaries, n.d.). This definition is distinct to the public office and abuse of power definitions in two ways. First, it makes no mention of people in positions of power – unlike the Oxford Dictionary's first definition of corruption, which defines it as 'dishonest or fraudulent conduct by those in power, typically involving bribery' (Oxford Dictionaries, n.d.). Second, this definition makes no mention of personal gain. As academic writings highlight, the decay definition is distinct as it focuses on the moral atrophy of individuals and institutions. The individual side of the definition is apparent in Peter Bratsis' analysis of corruption in ancient Greece. He argues that the term *diaphtheirein*, the Greek equivalent for the word bribery, referred to 'the corruption of the mind by which the ability to make sound judgments and pursue the good has been impaired' (Bratsis, 2003: 12). Here corruption means to 'pervert, degrade, ruin and debase' (Williams, 1999: 504). In other words, bribery was thought about in terms of individual immorality.

Analysing historical narratives about corruption also reveals the institutional side of the decay definition. For example, Machiavelli associated the term corruption (*corruzione*) with the deterioration of the quality of government, regardless of the reasons (Bratsis, 2003). Similarly, the Enlightenment economist Adam Ferguson (1723–1816) wrote about corruption as a type of progress-induced decline of civic virtue and politics (Hill, 2012). More recently Mlada Bukovansky (2006) – drawing on republican theorists such as Quentin Skinner and Machiavelli – has argued that the emphasis that republican political thought places on institutional corruption provides a corrective to the limitations of liberal and rationalist discourses offered by scholars and development organisations.

From this perspective, corruption is also considered a term of condemnation towards acts that are denounced by a particular community. The types of acts that are covered by this definition are culturally determined depending 'on the context and on the position of the actors involved' (de Sardan 1999: 34). From this perspective, corruption comes into being when immoral acts benefit individuals or communities outside of the group (de Sardan, 1999), or when individuals within the community transgress communal norms for individual benefit (Skinner, 1990; Williams, 1999). Thus, corruption is a label applied to activities that a group of people disagrees with; it is not applied to activity that the group itself considers acceptable.

When discussing causes of corruption, those who reflect the moral perspective focus on processes that undermine 'traditional' and/or 'communal' values. For example, de Sardan (1999) argues that corruption occurs when shaming, a cultural mechanism for keeping corruption in check in Africa, is no longer effective. Likewise, Williams suggests that corruption is marked by societies that discard 'civic virtue and social responsibility' in favour of 'intense competition for spoils' (Williams, 1999: 504). Such competition may be exacerbated by what de Sardan calls 'over-monetisation', where 'personal relationships ... take on a permanent monetary form' (de Sardan, 1999: 45). In Africa this includes:

> giving of 'taxi fare' to a visitor, giving coins to the children of friends, giving money for the purchase of a length of African print to a cousin going to a school party.
>
> (de Sardan, 1999: 45)

When relationships are maintained by money this 'obliges all and sundry to engage in a permanent quest for "means"'; that is, corruption becomes, for some in the community, a monetised extension of traditional gift giving (de Sardan, 1999: 46). So, for scholars who reflect the moral perspective, corruption is caused by the clash of traditional/communal and modern norms and expectations.

From this perspective, solutions to corruption should emerge from within individual communities. Richards calls these 'community-based anti-corruption programmes', which are 'initiatives that are physically and conceptually located in a community to fight and counter corruption' (Richards, 2006: 5). Such measures include community programmes to monitor corruption, media initiatives that shape opinions around corrupt practices, and school programmes that teach appropriate norms (Michael, 2004). They stress the importance of the involvement of 'civil society' and NGOs to address corruption. Quite often those approaching anti-corruption from the moral perspective believe existing large-scale transnational anti-corruption organisations and institutions could better fight corruption if they incorporated local understandings in their policies and programmes (de Sardan, 1999; Huffer, 2005; Bukovansky, 2006). From this perspective, local understandings can shape anti-corruption policies and programmes to ensure their relevance within the contexts in which they operate.

The *critical perspective on corruption* is particularly concerned with the effects of the anti-corruption industry. It is associated with authors who take aim at the industry for depoliticising corruption and responses to it (Brown & Cloke, 2006, 2011), reinforcing neo-liberalism (Hindess, 2005), failing to address corruption (Brown & Cloke, 2011) and reinforcing unequal relations between Western and local actors (Polzer, 2001; de Maria, 2005, 2008; Hindess, 2005; Harrison, 2006, 2007). There are two key strands of thought that have come to shape the critical perspective. First, neo-Marxist critics of (anti-)corruption are concerned with the ways in which the anti-corruption industry perpetuates the problems embedded in the capitalist commodity production process and the social relations woven around it (Hindess, 2005; Brown & Cloke, 2006, 2011; Rocha et al., 2011; Murphy, 2011). For example, Rocha and colleagues (2011) draw on the work of the neo-Marxist geographer

David Harvey to critique the way anti-corruption initiatives have strengthened class power in Nicaragua.

Second, there is a post-structuralist school (Michael, 2004), whose arguments resonate with those of post-development theorists (Escobar, 1984, 1992, 1995, 2003; Rahnema, 1992; Sachs, 1992; Ferguson, 1994). Drawing upon post-structuralist authors (particularly Michel Foucault), post-development theorists deconstruct 'development discourse', which they view as a '"narrative" of Western hegemony bent on denying or destroying popular practices and knowledge' (de Sardan, 2005: 5). They unmask the way development discourse has alienated local communities from the very projects and policies designed to improve their lives.

Some who write from a critical perspective mirror this post-structuralist approach. Polzer (2001), for instance, draws from Foucault, and references the most famous post-development theorist, Arturo Escobar, to critique the World Bank's anti-corruption policies. Those writing from this approach also have a tendency to view the anti-corruption industry as an extension of the development industry. Sampson (2010: 261) writes that the 'characteristics of the anti-corruption industry, including anti-corruptionist discourse, resemble that which has taken place in development aid'. Although some have tried to distance themselves from this contentious body of work (Polzer, 2001; Harrison, 2006), many writing from the critical perspective reflect concerns raised by post-development theorists, the key difference being the focus upon anti-corruption activities, a part of the development industry that has been largely overlooked by post-development scholarship.

Despite differences in their theoretical foundations, in many ways the neo-Marxist and post-structuralist critiques share similar views about the way corruption is defined, causes ascribed, and the types of solutions that should be prescribed. Unlike the previous perspectives, the critical perspective on corruption does not seek to define corruption per se; rather it highlights how definitions used by anti-corruption organisations marginalise competing understandings. For example, Harrison (2006: 21) argues that the 'ability to define an act as corrupt or to partake in "corrupt" activity reflects power, something that is seldom made explicit in international anti-corruption rhetoric'. So, when anti-corruption organisations suggest that one country is more or less corrupt than another, based upon their definition (or interpretation) of corruption, critical scholars suggest that this reveals the Western-centric bias of the organisations themselves, as much as, if not more than, it shows anything about where corruption is manifest (Hindess, 2005; Harrison, 2006; Brown & Cloke, 2011). However, as previously described this approach has arguably been influential in broadening mainstream definitions of corruption.

The critical perspective focuses on not only the way the anti-corruption 'industry' causes corruption, but also how it leads to other negative outcomes. For example, Anechiarico and Jacobs (1996) highlight the failure of anti-corruption programmes in New York, suggesting that they reduced the capacity of the public administration to monitor corruption, and resulted in corruption and inefficiencies far worse than those which these programmes set out to prevent. For others, the anti-corruption industry's solutions have transferred the problem of corruption to the private sector (de Maria, 2005; Holmes, 2006b), reinforcing unequal relations between developed and developing nations (Jayasuriya, 2002). These concerns are linked by an overarching concern about power relations. Critical scholars want to know how practices and discourses on corruption shape and reinforce power relations.

Many evoking the critical perspective valorise solutions that stem from grassroots groups that are 'truly independent and intrinsically motivated' (Moroff & Schmidt-Pfister, 2010: 93). As a corollary, some who are critical of anti-corruption organisations fear that 'authentic' local responses will be swept away in the maelstrom of the global anti-corruption agenda. Sampson (2010: 269) argues that the 'major players' (donors, governments and private NGOs) of the development industry marginalise those who question their conventional approach, thereby limiting grassroots input. What is more concerning for Sampson and others (Krastev, 2004; Di Puppo, 2010; Moroff & Schmidt-Pfister, 2010), however, is the tendency for grassroots movements to professionalise and mimic the activities of more established and better resourced organisations. Di Puppo's (2010) analysis of anti-corruption organisations shows that local actors working to mitigate corruption in Georgia are more accountable to foreign donors than to their domestic constituency.

In addition, some critical scholars argue that anti-corruption efforts need to be re-politicised (Huffer, 2005; Hindess, 2005; Brown & Cloke, 2011). For these scholars, the aid industry provides overly technical solutions to corruption that fail to critically address the corruption of private and public sector elites. As a result, they argue that corruption must be addressed through direct political action that confronts elites in both the state and the private sector. Hindess (2005), for example, is critical of TI and other agencies who view elites (particularly from the private sector) as anti-corruption champions, rather than what he believes they really are – key purveyors of corruption. In turn, he argues that international anti-corruption organisations are playing along to a neo-liberal script, by believing too heavily in the anti-state sentiments contained in the economic perspective.

As the name suggests, the mainstream view is more frequently represented in the literature than the alternative view. In turn the legal, public office and economic perspectives dominate the anti-corruption industry. These perspectives view corruption as a universal problem and thus allow for global responses to be applied at the local level. Challenging these mainstream perspectives are the moral and critical perspectives, which focus on the importance of local interpretations and responses to corruption. While still in the shadow of the mainstream view, the alternative view, as Michael (2004) and Brown and Cloke (2011) identify, has rapidly expanded over the past decade as scholars have become increasingly attuned to the limitations of mainstream approaches to addressing corruption. The way these perspectives are reflected in PNG is examined next.

Alternative and mainstream views about corruption in PNG

There is a growing literature about corruption in PNG, but little analysis of how the term 'corruption' is employed and understood. Drawing on the taxonomy developed in the previous section, this section analyses the way scholars have conceptualised corruption in PNG.

The mainstream view

The three perspectives of the mainstream view – legal, public office and economic – are reflected in scholarship and policy documents about PNG, but to varying degrees. Scholarship on PNG sometimes takes a *legal perspective on corruption*, with authors assuming that corruption is a subset or an extension of criminal activity (Pitts, 2001, 2002; Fallon, Sugden, & Pieper, 2003). This has led to the terms 'corruption' and 'crime' being used interchangeably. For example, throughout Maxine Pitts' (2002) book *Crime, Corruption and Capacity in Papua New Guinea,* corruption is assumed to be a criminal act, and there is little attempt to explain how corruption is different from illegal activity.

Many also express concern that PNG's judicial system has failed to adequately curtail corruption. The country has signed the United Nations Convention against Corruption (UNCAC). It also has a number of national laws that can be drawn on to fight corruption. These include the Organic Law on the Duties and Responsibilities of Leaders 1975 (which outlines the Leadership Code), Criminal Code Act 1974, Proceeds of Crime Act (POCA) 2005, Audit Act 1989 and the Electoral Reform Act 2007. But enforcing these laws is another matter entirely. Over a decade ago, Pitts (2002: 172) lamented that there has been much discussion in PNG about bringing those guilty of corruption to justice, but 'beyond Commissions of Inquiry, few prosecutions ensue, and even fewer perpetrators are punished, because under-resourced agencies either fail to apprehend or to prosecute criminals'. While there have been some successes, the situation by and large remains the same to this day. Many believe that the laws themselves are adequate although poorly enforced (Dix & Pok, 2009; Minta, 2014), and there have been numerous calls for stricter enforcement of existing laws, particularly for high-profile politicians and public servants, in order to send a message to citizens that corruption will not be tolerated.

While PNG has a number of existing laws to combat corruption, there are – from time to time – calls for new laws. For instance, *The Commission of Inquiry Generally Into The Department of Finance* (Davani, Sheehan, & Manoa, 2009) found numerous examples of corruption in the PNG Department of Finance; it made 75 recommendations to reduce mismanagement and corruption both in the department and across the larger bureaucracy. Some recommendations called for new laws, such as:

- a Whistle Blower Protection Act … to provide legal protection for persons and public officers who report corrupt practices by public officials;

- a Freedom of Information Act to provide clear processes to regulate access to official records and documents;
- Legislation … that promotes the use of assessors in criminal trials preparatory to eventual adoption of a system of jury trials for major crime. (Davani et al., 2009: 797–798)

These laws, the inquiry believed, would help address corruption in bureaucracies across the country.

Legal institutions have also failed to make significant impacts on corruption. As explored in chapter 4, the country's constitutionally mandated OC PNG has been described as under-resourced and underperforming (Okole & Kavanamur, 2003; Justice Advisory Group, 2005). As a result, some argue for the introduction of additional legal institutions, such as an Independent Commission Against Corruption (ICAC) with a wider range of investigatory powers. A draft Organic Law for an ICAC was proposed for PNG in 1997 and found support among anti-corruption organisations and academics (Mana, 1999). However, it failed to garner enough political support and was not passed by Parliament. PNG has more recently had another attempt at introducing an ICAC. In 2011 the government developed draft legislation to establish an ICAC in 2013. In November 2015 the PNG Parliament passed legislation enabling the anti-corruption agency to be established. At the time of writing the administrative structure was being determined by a commissioner and the agency had yet to be fully established.

Mirroring the *public office perspective on corruption*, the most popular perspective taken by those writing about corruption in PNG, many are concerned that Papua New Guinean state officials fail to adequately follow due process (Dinnen, 1997; Dorney, 2001; Hughes, 2003; Gosarevski et al., 2004b). Their concerns appear well-founded, with investigations into the public sector often finding examples of gross mismanagement and corruption. For instance, in a report into 20 government agencies, the Auditor General found many instances of poor management practices, including:

- Asset registers that were either non-existent or were not maintained properly in all 20 agencies;
- Periodic stock-takes not being conducted to determine the accuracy of assets on hand in 16 agencies;
- Lack of records of portable and attractive items, i.e. mobile phones and laptops;
- Non-existent controls surrounding management of vehicle fleets (Office of the Auditor-General of Papua New Guinea, 2010: 6).

Similarly, in a damning review, the Public Accounts Committee described the government's public accounting system as 'weak, corrupt and in direct breach of the Public Finances Management Act' (ABC Radio Australia, 2011). Such findings highlight the ongoing concern that corruption is caused by the inability, or unwillingness, of state officials to comply with rules and regulations.

It is often argued that PNG's cultural and political fragmentation is a key reason for state officials transgressing rules and regulations. In PNG there is a tendency for citizens to identify with small-scale communities that are often divided along ethno-linguistic lines. In addition, there have been numerous 'micronationalist' movements – disparate groups united by their predilection towards 'disengagement or withdrawal (but not, as a rule, formal secession) from the larger, national community' (May, 2004: 49). Calls from Bougainville province and the Papuan separatist movement for secession from PNG are two of many such movements that have sprung up around the country since independence. In turn, writings about PNG stress the 'inchoate, developing character' of the nation (Robbins, 1998: 108); PNG appears to many as 'an unimagined community, a failed imitation of more established, more homogeneous Western nations' (Foster, 2002: 3). As a result, the nation-state is often subverted (Dinnen, 2001) or ignored by many in favour of sub-national affiliations.

This has consequences for practices and perceptions of corruption. According to Payani (2000), Papua New Guineans' tendency to identify with clan and community, rather than the state, has resulted in widespread nepotism in the bureaucracy. Through participant observation of administrative offices in four central government departments, Payani finds that the bureaucracy has become 'ethnicised', with 'departments and agencies [dominated] by persons from a particular tribe, province, or region at the expense of others' (Payani, 2000: 143). As a result, Payani believes segments of the state have become an extension of communities who use state power to benefit their friends, family and community. Gordon and Meggitt (1985: 181) explain how this system operates: 'once a successful politician, public servant, or entrepreneur establishes ... a bridgehead, other members of his clan exploit the entry and insert themselves in the administrative machinery.' So, the allegiance between state officials and their kith and kin is seen to enable patronage networks that facilitate corruption.

Patronage networks have also, it is argued, shaped the democratic process in PNG. Many fear that the nation has become a clientalistic political system where patrons (political leaders) provide material benefits to clients (voters/ followers) in return for political support (Kurer, 2007; Standish, 2007; Allen & Hasnain, 2010). Indeed, it is feared that the political system has come to revolve around personalities rather than parties, with charismatic politicians elected by local constituents who expect monetary rewards for their support (Kurer, 2007; Standish, 2007). There have been numerous reports about candidates distributing increasingly large sums of money and gifts to secure votes in their own village, clan or tribal grouping (Stewart & Strathern, 1998; Ketan, 2004, 2007; May, 2006; Standish, 2007). The practice has become so widespread that some believe that, for their part, citizens actively encourage this behaviour. For example, May (2004: 320) suggests that many Papua New Guineans seek 'additional compensation for land purchased earlier by the government for schools, airstrips, roads, towns, and other public facilities, and often threaten violent action if their demands are not met'. In this way, it is argued, corruption has spread, contagion-like, throughout PNG.

There has been an influential, albeit smaller group of academics writing about corruption in PNG from an *economic perspective.* For example, Hughes suggested the government of PNG should not rely on income from large-scale development projects because these projects 'create ... economic rents that provide revenues for a swollen government and public services' that fuel corruption (Hughes, 2004: 2). As a corollary, some argue that reducing the PNG government's hold over resource rents, while creating conditions for greater private sector investment, could address corruption in PNG. For example, Gosarevski et al. (2004a) believe that reducing unnecessary state regulation and opening up the economy to increased foreign direct investment deprives politicians and bureaucrats of the rents that enable corruption. It would, they argue, also result in greater opposition to corrupt transactions, implying that the private sector is less prone to corruption than the public sector.

The alternative view

The strongest advocates for the alternative view have been mostly indigenous scholars and policy makers, who have reflected the *moral perspective.* For example, Joe Kanekane (2007) – at one time the president of the PNG Media Council – suggests that Papua New Guinean customary ways result in a tolerance of corruption. He explains that voters re-elect corrupt leaders as they have a 'primordial acceptance that, although MPs have done wrong, they should be given a second chance' (Kanekane, 2007: 23). Papua New Guineans, according to Kanekane, place personal relationships before rules and laws of the state, making corruption a concept that is somewhat foreign to local culture and traditions.

Local academics have also argued for local solutions to corruption. Okole and Kavanamur (2003) argue that 'the dynamism and fluidity of the *wantok* system [a local system of obligation between kith and kin], if it can be elevated ... to the level of a civil society-wide network, would be invaluable to stem the flow of corruption' (Okole & Kavanamur, 2003: 24). In other words, they regard cultural systems as strengths to address corruption. Some international academics have been sympathetic to the moral perspective. For example, Crocombe (2001) suggests that while citizens may not approve of corruption when considered abstractly, they may condone 'corrupt' acts and rationalise them if they benefit from them. This sympathetic view suggests that corruption may be a weapon of the weak in response to poverty and inequality.

Those writing from the *critical perspective* have focused on the Australian aid programme (previously known as the Australian Agency for International Development, AusAID, and now operated through the Department of Foreign Affairs and Trade, DFAT) – PNG's largest donor – and its policies and programmes. For example, Temby (2007: 38) draws upon the work of

post-development scholars to argue that AusAID's approach to good governance sidelines 'alternative and possibly more appropriate responses to the
challenges facing PNG'. A few have also accused the aid programme of
engaging in or perpetuating corruption. Tim Anderson, an Australian-based
academic, argues that given the large portions of aid monies returning to
Australia through 'boomerang aid' (aid money that returns to its country of
origin as it mostly benefits organisations and individuals of the donor nation
[Aid Watch, 2005]), corruption is a term that should be applied to the aid
programme itself (*PNG Post-Courier*, 2003). Others have suggested that
AusAID's approach to development, at least in the past, has left open the
potential for others to engage in corruption (Feeny, 2005; Temby, 2007). Feeny
(2005), for instance, argues that AusAID shifted its approach to delivering aid
money to PNG, from budget support to programmatic support, because funds
provided under the previous approach were being misappropriated.

While some have expressed concern over AusAID's good governance programme, there has been scant critique of anti-corruption NGOs and other
civil society groups. In a short article, Larmour (2003) provides some insight
into (rather than critique of) the operations of TI PNG. Larmour (2003)
acknowledges that the impacts of TI PNG's policies are difficult to assess, but
is still concerned that a number of its high-profile advocacy campaigns were
not sufficiently supported by politicians and the broader public. In his 2012
book, Larmour also highlights some of the disjunctures between cultural
interpretations of corruption and TI PNG's attempts to address it (Larmour,
2012). In so doing he raises questions about the 'fit' between TI PNG's
anti-corruption strategies and local people's priorities.

Conclusion

To capture the diversity of views about corruption in the country this chapter has
introduced a theoretical framework on corruption based on the literature. The
literature on corruption in PNG reflects this framework, and suggests that
there is a dearth of scholarship on the alternative view on corruption, despite the
importance of structural concerns – poverty, culture and poor social indicators
(see chapter 1) – that suggest further exploration of these perspectives could be
fruitful in understanding the nature of corruption in the country. Future chapters
will draw upon this framework and analyse how citizens and anti-corruption
actors reflect the mainstream and alternative views. But first, the following
chapter highlights the multiple scales through which corruption plays out in
PNG. It shows that understanding the veracity of these perspectives in PNG
requires paying close attention to the scales where corruption plays out.

Note

1 Huffer's (2005) adroit analysis also draws on Williams (1999), but does not set out
   the type of framework to analyse corruption that this chapter does.

3 Corruption in PNG

Perspectives and evidence

Introduction

In much of the international literature corruption is presented as a national problem. This is perpetuated by international rankings, such as Transparency International's (TI) Corruption Perceptions Index (CPI) and the World Bank's World Governance Indicators. These indicators rank individual countries by perceived levels of corruption. However, understanding corruption requires researchers and policy makers to move beyond methodologies that fetishise the nation-state. Corruption is embedded in and connected to a myriad of spatial scales, from local through to national and international scales.

This chapter draws on findings from a large-scale survey, focus groups and an analysis of national and international narratives and scandals. It highlights the different scales at which corruption is manifest, caused and interpreted within and beyond the nation-state of PNG. This is a particularly important approach given the weakness of the PNG state. The PNG state is often only weakly felt in rural and remote parts of the country, where most citizens live (Walton, 2015). The state also struggles to secure its borders. Incursions across Indonesia and PNG's land border are frequent, and there is little monitoring of the country's sea borders. This has produced concern among some Australian analysts who fear that the poor policing of Australia's northern border with PNG – through the Torres Strait – makes it vulnerable to illegal trafficking and the spread of multi-drug resistant tuberculosis (TB) (Sharp, 2016).

To understand corruption in PNG as a multi-scalar phenomenon the chapter first examines the way customs, socio-political and economic factors shape attitudes towards and responses to corruption at the local level. Next, it examines the national scale, and the institutionalisation of corruption within the state. Then, it focuses on the way the international system facilitates transnational flows of illicit corruption from PNG. The conclusion reflects on what the multi-scalar nature of corruption means for understanding and addressing corruption in PNG.

Local perspectives on corruption

Most of PNG's citizens live in the shadow of the state. Approximately 80 per cent of the population is dependent on subsistence agriculture, with around the same percentage of the country's 7.5 million citizens living in rural areas. In these areas the state's presence is particularly weak, and in some ways is getting weaker. A review of the state of schools and health facilities across 8 out of 22 provinces between 2002 and 2012 (Howes et al., 2014) showed that while primary schools have significantly improved, health facilities – including small aid posts that service remote locations – have worsened. Tragically, researchers (including the author) found that there were fewer orderlies, fewer available drugs, and facilities were in a worse state of repair in 2012 compared to 2002. Conditions were worst in remote locations. Given this, it is not surprising that PNG has had poor health indicators for more than a decade.

The importance of state weakness and deterioration for citizens' attitudes towards corruption is apparent when examining data from focus group discussions and a household survey conducted across the country. Conducted with TI PNG, the research aimed to better understand the ways in which Papua New Guineans interpreted corruption, and what this might mean for anti-corruption responses. Researchers spoke to over 2,300 people and travelled to some of the most remote corners of the country. Results showed that state weakness plays a role in determining how corruption (or *korapsen* in Tok Pisin) is understood.

Citizens' definitions of corruption in PNG

The state is central to the mainstream view on corruption but what happens to people's understanding of corruption when the state is weak and, in some cases, seemingly non-existent? To get a sense of the nature of state weakness in determining perspectives on corruption I conducted focus group discussions in remote locations in four provinces – Southern Highlands, Madang, Milne Bay, and East New Britain – of the country. Respondents were located in two villages in each province – one with good access to resources and the other far away from them, in areas where the state was by and large absent. A total of 80 focus groups were conducted with men and women, the young and the old and community leaders.

In focus groups respondents were asked to define corruption and provide examples of what it might be. Most respondents knew what corruption, or *korapsen* in the local lingua franca, Tok Pisin, was but there were a few difficulties around translation. A few marginalised groups – particularly those with little or no education – had never heard of the word *korapsen*/corruption (because the words are homonyms it was impossible to know which word respondents were responding to). This meant that researchers – who were either from or working with the communities involved and familiar with the local language – occasionally had to translate this term into the local language. This sometimes required a degree of interpretation. I had anticipated this, and designed scenarios to allow Papua New Guineans from all walks of life to give their thoughts on what they believed to be corrupt. We'll get to the approach and analysis of that research shortly. But before doing so it is worthwhile examining responses from those (a majority) who did understand the concept.

In line with the discussion of different definitions of corruption in chapter 2, respondents defined corruption in three ways. First, respondents spoke of corruption in terms of the public office definition of corruption that equates corrupt conduct to that which involves state officials. This included, in the words of a male respondent from the coastal province of Madang, 'breaking government law, leaders misleading people, public servants misusing government vehicles' and other acts involving state officials misusing their position. Respondents relayed stories about politicians or local leaders making pornography, getting involved with prostitution or leaving their wives for younger women. For example, a woman in Southern Highlands Province said, 'corruption is ... our leaders going to town and sleeping with women ... drinking beer ... and leaving their wives.' For many, becoming a leader or politician was not simply about upholding laws or regulations, it was also about upholding moral standards – a point central to the country's Leadership Code, a legal code of conduct for state officials (see chapter 4). These were common complaints that were echoed across the different villages visited. However, the more remote the places, the more likely it was that alternative definitions of corruption appeared.

In places where non-state institutions held sway, respondents were more likely to reflect the 'abuse of power' definition which broadens out the types of actors involved in corruption. For example, in Madang many equated corruption with the operations of the nearby business Ramu Agri-Industries Limited. They suggested that this company engaged in nepotism, with employees hired based upon their personal connections rather than merit. There were also suggestions that private sector organisations were corrupt due to excessive rent seeking. For a man in his twenties in Madang, corruption was about 'businessmen thinking of making profits'. Non-profit organisations, like the church, were also seen to fall under the umbrella of corruption. In the island province of East New Britain a few respondents suggested, in hushed tones, that the local clergy embezzling church funds was an example of corruption. These examples suggest that corruption was seen as more than the abuse of public office; rather, for these respondents it fit into the 'abuse of entrusted power for private gain' definition.

Among more marginalised groups – particularly those who were uneducated and in very remote locations – corruption meant something altogether different. For these respondents corruption had little to do with those entrusted with power by the state. Rather corruption was tied to small-scale immoral acts that implicated citizens. There were many who reflected the *individual* decay definition. This included lying, gossip, prostitution, womanising, making and drinking homebrew and other morally dubious activities involving citizens. For women in the Southern Highlands corruption was personal. It meant 'arguments and fights', 'drunk people in public places like the market', 'bad thoughts, jealousy', 'rape cases' and 'disrespect, adultery, no love, no homes and loss of life'. These were not outcomes of corruption; they were inextricably tied up with the meaning of the term. Whether the perpetrators of these acts held legitimised positions of power was, by and large, not the issue. Indeed, only a few women in Southern Highlands connected the term corruption to leaders. The threat of violent men – whether they were brothers, husbands or strangers – was what they were mostly concerned about, regardless of the position they held.

Respondents also equated corruption to *institutional* decay. In Southern Highlands Province, respondents spoke of corruption as poor government standards, while those in Milne Bay considered corruption as the lack of government services in their area. In Madang, one respondent said corruption occurs when the *wantok* system benefits individuals more than the community. Notwithstanding these examples, most respondents reflected the individual side of the decay definition (which is why the example of decay in the scenarios reflects the individual side of the definition, more on this later).

To get a sense as to how representative these different definitions were, a follow-up survey was commissioned, again with the involvement of TI PNG. The survey was carried out across nine provinces of the country and included 1,825 respondents.

The survey included a series of scenarios to represent different types, scales and definitions of corruption (see Table 3.1). After the scenarios were presented (verbally and in pictorial form), respondents were asked how 'corrupt', 'acceptable' and 'harmful' they believed the scenarios to be. The scenarios and questions were derived from studies into citizen and elite perceptions of corruption (Peters & Welch, 1978; Johnston, 1986; Independent Commission Against Corruption, 1994; Jackson & Smith, 1996), pre-tested and adapted for the local context.

As Table 3.1 highlights, each scenario represented a different scale, definition, type and locale. Out of the nine scenarios, five scenarios reflected petty, or small-scale, types of corruption; that is the scenarios suggest that the monetary value of the transaction or activity was relatively small. The scenarios also reflected the different definitions of corruption: public office, abuse of power and the decay definition (which included a scenario involving individual decay rather than institutional decay). Different types of corruption were also reflected, including bribery, undue influence, nepotism, embezzlement, conflict of interest and immoral behaviour. Six of the scenarios were more likely to be acts that played out where citizens lived (labelled 'local'), while three were classified as 'far away' and likely to occur outside of the purview of concerned citizens.

The results can surprise those uninitiated with PNG. During a lecture for a Masters course on corruption and anti-corruption in Canberra, Australia, I asked students – who came from countries all around the world, but not from PNG – to nominate which scenarios they believed would be considered totally corrupt by Papua New Guineans. The students were provided with a rundown – admittedly brief – on the economic, political and cultural factors that might

Table 3.1 Scenarios presented to respondents

Scenario	Code	Scale	Definition	Type	Locale
A contractor hands money to a public servant in order to be favoured in a contract bid	CONTRACTOR	Large[1]	Public office	Bribery	Far away
A voter accepts an offer to sell his vote to a candidate for 50 kina	VOTER	Petty	Public office	Bribery	Local
A logging company gets logging access to customary land by flying customary leaders to Australia and giving them gifts, without consultation with other community members	LOGGING COMPANY	Large	Abuse of power	Undue influence	Local
After a large company legally influences politicians, the government passes a law which helps them make greater profits	LARGE COMPANY	Large	Abuse of power	Undue influence	Far away
A man is employed as a driver for a government department by his *wantok* [relation/friend] without going through a recruitment process. He is a safe and reliable driver	DRIVER	Petty	Public office	Nepotism	Local
A teacher takes pens and note pads from her school stores cupboard to use for her church meetings	TEACHER	Petty	Public office	Embezzlement	Local
Electoral workers are provided with food and drink by a candidate	ELECTORAL WORKER	Petty	Public office	Undue influence	Local
A Minister for Defence owns a company with which the Defence Department has a million dollar contract	MINISTER OF DEFENCE	Large	Public office	Conflict of interest	Far away
A young woman is drinking homebrew [homemade alcohol] and selling sex	HOMEBREW	Petty	Decay[2]	Unethical behaviour	Local

shape perceptions of corruption in the country. They were also well versed in the complexity of corruption and the variety of ways it can be defined. Despite this there were gasps of surprise when I presented survey findings and revealed how students' responses differed from Papua New Guineans'. For example, while students were strongly convinced of the corruption of the Contractor scenario, Papua New Guinean respondents were not (Figure 3.1). Likewise, while students did not put up their hands when asked if the Homebrew scenario was corrupt, Figure 3.1 reveals it was, for respondents, the most corrupt, most unacceptable and most harmful scenario of the nine. These differences came about, in part, due to Papua New Guinean respondents being shaped by broader definitions of corruption and the local impacts of corrupt activity.

When we directly compare the scenarios in terms of the definition they represent, little separated the public office and abuse of power definitions, particularly when it came to assessing the scenarios in terms of their degree of unacceptability and corruption. This is evident in Figure 3.2, which compares the decay definition (represented by the Homebrew scenario), the public office definition (an average of the Contractor, Voter, Driver, Teacher, Electoral Worker and Minister of Defence scenarios) and the abuse of power definition (an average of the Logging Company and Large Company scenarios).

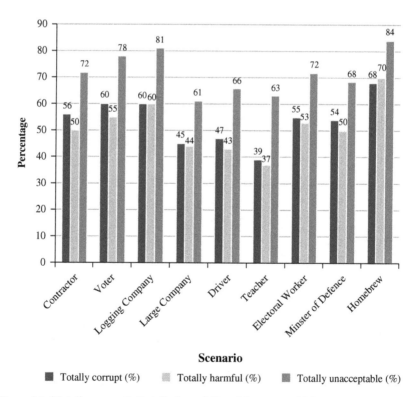

Figure 3.1 'Totally corrupt', 'totally harmful' and 'unacceptable' responses

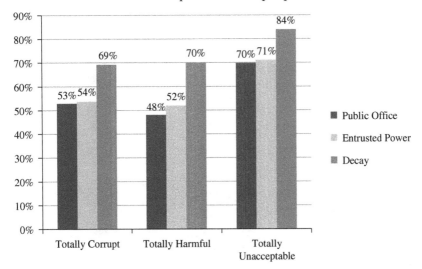

Figure 3.2 Comparison of responses of 'totally corrupt', 'totally harmful' and 'unacceptable' by definition

What variables shaped these responses? Figure 3.3 shows that the size of corruption really didn't matter that much for respondents. When categorising the different definitions of corruption, in terms of their perceived level of harm, corruption and unacceptability, only a few percentage points separated scenarios depicting small- (petty) and large-scale corruption.

What was far more important for responses to these scenarios was the likelihood that they would play out within a local context, which meant that citizens were more likely to be directly impacted by them. Figure 3.4 shows that respondents were more likely to express concern about the harm, corruption and unacceptability of scenarios likely to play out at the local scale. In particular, respondents expressed much greater concern about the level of harm that local scenarios would cause. This explains the difference between the students' and Papua New Guinean responses. Students were well attuned to the harm that secretive corruption involving government and private sector elites could cause within places like PNG. Respondents were less concerned.

The importance of local concerns is particularly apparent when examining responses to scenarios representing different definitions of corruption. The scenario representing the abuse of power definition (Logging Company) was considered more corrupt by men (66 per cent compared to 54 per cent of women) and those with higher levels of education (57 per cent of those with education up to year nine considered this scenario corrupt compared to 71 per cent of those with higher education). The Voter scenario (representing the public office definition) was most likely considered unacceptable by urbanites (69 per cent compared to 57 per cent of rural dwellers), men (64 per cent compared to 56 per cent of women) and those with higher levels of education

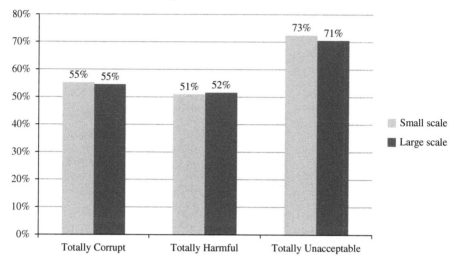

Figure 3.3 Comparison of responses of 'totally corrupt', 'totally harmful' and 'unacceptable' by scale of corruption

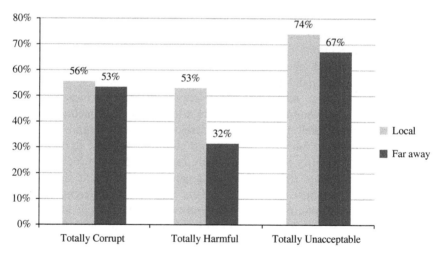

Figure 3.4 Comparison of responses of 'totally corrupt', 'totally harmful' and 'unacceptable' by 'localness'

(73 per cent compared to 56 per cent of those with education up to year nine). The scenario representing the decay definition, the Homebrew scenario, was likely to be considered unacceptable by those on lower incomes (73 per cent of those earning under 500 kina per month, compared to 58 per cent earning above 500 kina) and by women (72 per cent compared to 67 per cent of men).

As I have argued elsewhere (Walton, 2015), those who are more enfranchised within society are more likely to reflect the public office definition. This is likely due to these groups better understanding the rules and laws of the state. Those in urban areas with higher incomes have better access to the media, which frequently run stories about corruption and other illegal activities. Education facilitates awareness about how the electoral system should function, as well as citizens' rights and responsibilities. Given that women can embody marginalisation in PNG, it is not surprising that women were far less likely to consider the Voter scenario corrupt than males. In PNG, women's marginalisation mirrors that of the poor. According to the 2000 Census, women in PNG had lower education, literacy and wage employment rates than males (Rannells & Matatier, 2005: 247). They are also poorly represented in politics – the 2012 elections saw 3 women elected to the National Parliament (out of a possible 111 seats), which, along with the 1977 election, was the highest number of women ever elected to Parliament since the first general election in 1964 (Sepoe, 2002; True et al., 2012).[3] In PNG marginalisation is a product of low income and, more broadly, as championed by Amartya Sen (1999), the denial of social, economic and political capabilities. Both men and women are denied the opportunities to fulfil their capabilities, but in PNG it is women who are most often marginalised.

Marginalisation shaped responses to the scenarios in the focus groups. Respondents, particularly women and those in poorer villages, said that the Voter scenario could be beneficial if it enfranchised everyone in the community. For example, one man from Southern Highlands Province said the transaction should be welcomed if it benefits the entire community. A woman from Southern Highlands Province said: 'the candidate is to be blamed as he is doing it secretly, which is not good. If he comes openly and lets the community leaders know, then that's good.' In the midst of poor state services and governance, respondents said that such a scenario is to be welcomed if everyone benefits and the transaction is carried out transparently. So, respondents unable to understand the laws and rules of the state, and those marginalised economically and socially, could justify the Voter scenario. Those who are less enfranchised have come to depend upon the illicit distribution of state benefits given the weakness of the state. This has in turn shaped their understandings of corruption.

Citizens' perceptions on the causes of corruption

In-depth discussions about some of these scenarios provided significant insights into respondents' perceptions about the causes of corruption. Respondents fell into two key groups: 'sympathetic' and 'unsympathetic'. Most respondents fell into the latter category; that is they were critical of the role of the protagonists and condemned the activities described. However, as the 'sympathetic' label suggests, there were a number of respondents who could understand why such activity would occur. Those sympathetic to the scenarios were more likely to be those without the wherewithal to benefit from state resources: those with little education, and women from remote locations.

Sympathetic responses were more prevalent with the *Wantok* and Voter scenarios in particular. An analysis of responses towards the former highlights the significant split between these two groups.

Respondents were unsympathetic towards Mary, the public servant in this scenario, because they believed she should have followed the correct procedures for employing her *wantok*. For example, a male in Madang said, 'Mary feels she has the right to employ her *wantok* without advertising the position, but she's wrong in doing that because she should follow the right procedure to recruit for any available position.' As a result, some suggested Mary should be penalised, with a female from Milne Bay saying, 'Mary should be suspended [from duty] for breaching the public service rule for recruitment.' There were some who also condemned Mary on moral grounds. For example, one male respondent from East New Britain said, 'Mary is practising nepotism as she is not honest in this issue.' Most were concerned that Mary failed to follow proper bureaucratic procedure.

Unsympathetic respondents also condemned Mary's *wantok*. Some argued that s/he must have known that Mary was not following correct procedure. For instance, one male from Southern Highlands said, 'Mary and her *wantok* are doing a corrupt thing, and they know they are [not following procedures].' Another male in the same focus group added, 'He [Mary's *wantok*] must have bribed her to get the job.' As a result some believed s/he should have refused the job; one male in Milne Bay, for instance, suggested, 'Mary's *wantok* should carefully refuse her offer.' For unsympathetic respondents – reflecting the public office perspective – both Mary and her *wantok* were at fault for failing to follow proper procedure.

Unsympathetic respondents blamed the *wantok* system for causing such a transaction. For example, a male in Waro village, Southern Highlands, said, 'Such things happen in PNG because of the *wantok* system [which is facilitated] through money or bribery.' For these respondents the *wantok* system unduly benefitted those who participated in it; in turn, they feared that participants accumulated social capital that could lead to future rewards. A male from a remote village in the Southern Highlands said, 'people practise the *wantok* system ... in order to be ranked as high profile in their own community.'

On the other hand, sympathetic respondents argued that Mary's actions were understandable, even philanthropic. A male from a remote village in Milne Bay suggested Mary was right to support her *wantok* as she was acting out of 'traditional obligations ... that means we can survive'. Some believed Mary was employing her *wantok* for protection. In a remote village in Madang, one woman said, 'women do not feel safe today therefore [Mary] needs someone to work with her to guarantee her security and safety.' For these respondents Mary's response was born out of traditional obligation and the need for protection.

This group also sympathised with Mary's *wantok* because they believed s/he was poor. One female from Milne Bay said, 'he accepted the offer because he needed the money'; another from Madang said, 'I think he needed the job so let him be.' As a result, some believed that Mary's actions could lift her *wantok* out of poverty. For example, one male from East New Britain argued that Mary's *wantok* 'could now eat rice'.

Sympathetic respondents believed scenarios such as this often occurred throughout PNG, and were caused by the difficulties of modern life. For example, a male from Madang said: 'life is hard so [we] should take whatever job comes by.' Others suggested that utilising friends and familial connections was particularly important when looking for jobs in a tight job market. For example, reflecting on the causes of such scenarios, one male from Waro, Southern Highlands, lamented that 'such things happen in PNG because a lot of family members do not have any job at all'. As a result, a male from Milne Bay said that, because it is hard to get jobs in PNG, the practice of 'who you know [nepotism] is the best solution'. Thus, these respondents viewed the *wantok* system as a crucial social protection system.

In sum, focus groups revealed some of the deep ambiguities in the narratives about corruption in PNG. Sympathetic respondents reflected the concerns of the alternative view on corruption by highlighting the structural constraints framing corruption. The mainstream view was reflected by those unsympathetic to the scenarios. While a majority of the focus groups reflected the mainstream view, the alternative view was particularly prevalent amongst those who were poor and politically marginalised (Walton, 2013b).

To get a sense of what people thought were the main drivers of corruption, the quantitative nine-province survey asked respondents to rate the degree of seriousness of seven possible causes. Table 3.2 shows that more respondents considered a lack of law enforcement and poor leadership to be the most serious causes. These drivers of corruption were more of a concern for respondents than the other issues that often get blamed for PNG's governance issues: poor salaries, the flawed electoral system, collusion between big business and government, lack of information and immorality. These responses, in part, reflect the areas where the state has been at its worst in PNG.

Table 3.2 Responses to statements about the causes of corruption in PNG

Cause	Serious (%)	Not serious (%)	Neither (%)
The morals of people are weak	41	49	11
Existing laws aren't enforced	65	22	11
The electoral system is flawed	38	46	16
Business pays for influence with government	28	53	20
Leadership is of a poor quality	64	24	12
Low salaries	27	59	14
Grassroots lack info on government spending	40	47	13

Law enforcement has been deteriorating since the country's independence, which is particularly apparent when examining the strength of the nation's police force, the Royal PNG Constabulary (RPNGC). The RPNGC has seen its ranks significantly decline: Dinnen, McLeod, and Peake (2006: 89) calculate that the ratio of police to the population at independence was 1:380, but more recent calculations place it at around 1:1,404 (Walton & Peiffer, 2015), well below the United Nations' recommended ratio of 1:450. The vastly male-dominated (male officers account for around 95 per cent) RPNGC has also struggled to undertake routine criminal investigations or apprehend suspects; moreover, management is often poor (Dinnen et al., 2006). These problems impede efforts to investigate and prosecute cases of corruption and are well understood by citizens throughout the country who are aware of the lack of impartial law enforcement.

Law enforcement was of particular concern to those respondents most likely to benefit from a functioning state system. Analysis of the quantitative survey revealed that 74 per cent of those with a post-high school qualification were concerned about law enforcement compared to only two-thirds of those with no or basic education. Similarly, two-thirds of those who earned below 100 kina per month were concerned about inadequate enforcement of the law compared to 83 per cent of those who earned over 2,000 kina per month (Walton, 2015).

The second major cause of corruption cited by respondents to the quantitative survey was poor leadership. The popularity of this response is also to be expected given the multiple instances of corruption involving PNG's leaders. Respondents were likely concerned with leaders at all scales including national political leaders and those from the community. Concern about poor leadership was evenly distributed between men and women, and those on different levels of income and education.

These responses suggest that many in PNG feel that the factors that drive corruption are – to a large extent – outside of their control. Decisions about policing are carried out in towns and cities far away from where most Papua New Guineans live. Moreover, few felt that they had control over their political leaders. Respondents in focus groups explained that if a political candidate was successful, and got into office, villagers would be quickly forgotten, justifying taking money from candidates as a short-term strategy of getting something from the election process. As a young male from Milne Bay said:

> The reason someone may take bribes is because as soon as these candidates win, they do not stick to their word and disappear into ... oblivion, leaving the voters to their own devices for the next five years ... so while the voters can they just take the cash and goods and gifts. So really, everyone does their own thing until it's time to repeat that cycle again.

The structural factors shaping corruption require broader and substantive nationwide responses that can better address poverty and inequalities. Poor law enforcement, for example, requires police and other state institutions that function properly. Meaningfully fixing such issues requires significant support from actors that can address these broader and more substantive structural issues, in line with the alternative view on corruption.

Fighting corruption in a weak state

Even if citizens came across an instance of corruption, the likelihood that they would report it was low. Only a quarter of respondents from the quantitative survey knew the process for reporting corruption. Similarly, few trusted state institutions to effectively respond to corruption. As Figure 3.5 demonstrates, key state institutions such as the Parliament, Office of the Prime Minister and police were considered totally or mostly effective at keeping government accountable by less than 30 per cent of respondents. The OC PNG was the only state institution to be considered effective in keeping government to account by (just) over half of respondents. Unsurprisingly, given the role they play in raising issues about corruption, NGOs and the churches – non-state actors – were considered the most effective. This suggests that these institutions are well regarded, but are they up to the task? Chapters 4 and 5 will examine the efforts of local and international NGOs to help answer this question.

This lack of trust in the ability of state actors to hold the government to account undermines citizens' willingness to report corruption to the authorities – even when they know how to report. My colleague Caryn Peiffer, from the University of Birmingham's Developmental Leadership Program, and I have examined the ways in which this trust impacts on citizens' willingness to report different types of corruption (Walton & Peiffer, 2015). Drawing on data generated from the nine-province quantitative survey, we looked at how

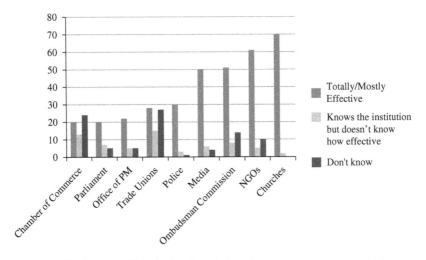

Figure 3.5 Effectiveness of key institutions in keeping government accountable

respondents' level of education, and their level of trust that the state would respond to corruption, impacted their willingness to report scenarios depicting corruption.

Using a series of statistical models, we found that formal education plays a crucial role in increasing citizens' willingness to report corruption to the authorities. Higher levels of *formal education* increased the likelihood of a respondent being willing to report by up to 31 per cent, and this relationship was significant across six out of eight scenarios[4] (only the Teacher and the Electoral Worker scenarios were not significant). Believing that nothing useful would be done about corruption reduced the likelihood of being willing to report by up to 43 per cent – this reduction was significant in three out of eight scenarios. However, when we used an interaction model, which examined the impact of education *and* state trust on reporting, the true impact of state weakness revealed itself.

The interaction model found that the belief that *something* would be done about corruption and higher levels of education resulted in an improvement in respondents' willingness to report. This improved the impact of formal education by up to 32 per cent. However, the inverse was also true. When people believed *nothing* would be done about corruption, education's positive impact reduced. In other words, when educated people didn't trust that the government would address corruption they were less likely to report it to the authorities.

In the context of PNG this is a very concerning finding. As further explained in the following section, the PNG state has very fragile anti-corruption agencies that have been undermined and attacked by political elites. Our findings suggest that these attacks have broader societal impacts; they are likely to reduce the willingness of educated citizens to report on corruption. This means that those best placed to fight corruption are less likely to do so.

Given respondents' belief that reporting corruption is ineffective, it is not surprising that many decide to take matters into their own hands. But the type of resistance depended upon the nature of corruption and the respondent's familiarity with it (or localness). Grouping responses to scenarios by their representation of grand and petty corruption (see Figure 3.6), we see that respondents said that they were more likely to report grand corruption to authorities, but much less likely to confront the perpetrators. More people also said they'd be likely to do nothing at all if faced with grand corruption rather than petty acts of corruption. In contrast, respondents said they'd more likely both directly confront and talk about petty corruption. This suggests that communities are more willing to personally challenge less powerful actors rather than powerful ones involved in large-scale corruption.

Figure 3.7 categorises responses by the likelihood that respondents would experience corruption 'locally' (as shown in Table 3.1, examples include the theft of classroom materials as a type of 'local corruption', and deals involving a senior government minister and a defence contract as a type of 'far away' corruption). It shows that the corruption experienced locally was more

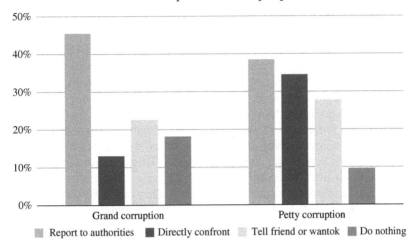

Figure 3.6 Types of responses to grand/petty corruption

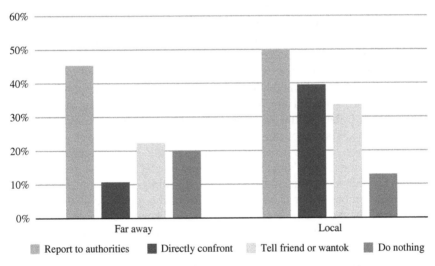

Figure 3.7 Types of responses to 'far away'/locally experienced corruption

likely to be acted on in terms of reporting the incident, confronting those involved and simply telling someone about it.

In focus groups, respondents spoke openly about how they might respond if faced with different types of corruption that they thought would impact their community. When presented the scenario of a teacher taking school materials and redistributing them (the Teacher scenario), some became angry. In some places respondents believed such a scenario would lead to arguments, threats, physical violence and even death, with tensions heightened between parents and the teacher, the teacher and his superiors/

colleagues and the teacher and students. In Southern Highlands Province, in particular, respondents suggested they would directly confront the teacher. For example, one male in a remote village in the Southern Highlands said, 'we'll beat up anyone who does this in our place.' In other areas, passive resistance was seen as a way of responding to such an occurrence. In a remote village in Milne Bay, a male leader said he would take his child out of school and refuse to pay school fees, 'if a teacher is going to be careless and shows negligence with the use of school materials'. It is difficult to overstate the level of passion in these focus group discussions – many were keen to express their disdain for such acts in the strongest of terms. This demonstrated that many people in PNG are concerned and willing to act against transgressions they believed would affect their community directly. They indicated that when faced with such a scenario they would not simply passively stand by.

Some respondents, however, had given up hope that fighting corruption would do any good given the dysfunctional nature of the state. For those sympathetic to the scenarios, PNG's dysfunctional state, poverty and cultural obligations meant that supporting corruption could be logical, even necessary. For example, many suggested that scenarios reflecting possible petty corruption could help address poverty. One male from East New Britain said that the public official's relative employed in the *Wantok* scenario was likely poor and that getting a job, no matter what the circumstances, meant they would pull themselves out of poverty. Sympathetic respondents tended to believe the scenarios could more equitably redistribute state resources. Responding to the Voter scenario, a male from Southern Highlands summed up this position, saying, 'such bribes should be given to the whole community and leaders so everyone benefits and not only a few people.' Respondents also spoke about the importance of kinship ties. Responding to the *Wantok* scenario, a man from the Southern Highlands said the public official was right to employ her relative as she was acting out of 'traditional obligations', and rightly supporting a member of her tribe. Females, those who identified as illiterate and those from the most remote villages were most likely to sympathise with scenarios depicting petty corruption. Again, it was these groups who were least likely to understand state rules and laws, and expressed concern that they were excluded from state resources. This means that state dysfunction, and the structural challenges it creates, not only reduce the likelihood of reporting corruption, but can also mean that supporting corruption becomes rational.

The state and corruption

While weak, PNG's state still looms large in the imagination of many. And these imaginings often include stories of corruption associated with the country's political elite. I was reminded of this most dramatically in the middle of 2008, when research coordinator Ivan Jemen and I conducted some preliminary focus groups to test the questions and scenarios we were to employ for these groups. After we'd finished administering our draft questionnaire in Port Moresby we took a stroll with some local youths and chatted about life in the settlements, work (or lack of it) and politics. As we ascended up a hill overlooking the administrative suburb of Waigaini, we eventually got around to the topic of our research – corruption. I asked what they thought was at the root of corruption in PNG. At once three or four of the settlers pointed straight at Parliament House. Laughing, one of the boys said, 'there is the root, and the trunk, branches and leaves of corruption.'

Politics is indeed at the heart of narratives about corruption within and outside the country. Politicians have been linked to corruption in the forestry sector (Barnett, 1989; Australian Conservation Foundation & CELCoR, 2006; Laurance et al., 2011), the misuse of their growing constituency funds (Ketan, 2007; Pok, 2015; Raitano, 2015), cooperation with businesspeople to defraud the state (Pok, 2014) and a range of other corrupt activities. Corruption scandals involving politicians are an almost daily occurrence, and are featured prominently within the nation's traditional and social media. Both forms of media have done a good job of both uncovering corruption scandals and intensifying rumours about the nature of the threat.

The ability of the media to highlight corruption is highlighted by a widely publicised sting, which showed how the mechanics of transnational corruption works. In June 2015 lawyers in PNG were caught on film explaining how they could launder corrupt funds to Australia through raising phony legal invoices. The video – secretly filmed by the international NGO Global Witness – showed lawyers explaining how laundered funds could be collected by corrupt PNG politicians through Australian bank accounts or property investments. One of the lawyers featured (who described the best way to bribe politicians) was a former Queensland Crown prosecutor, Greg Sheppard. The video was promoted through a special edition of *Dateline*, a news programme on the Australian television station SBS. The investigation was also featured through the Fairfax group of newspapers, which owns a number of mastheads across Australia (McKenzie, Baker, & Garnaut, 2015).

Even though there is a lot written about corruption in the country, few have revealed the way it works within and beyond PNG with so much detail. This has been particularly difficult to do given the secrecy associated with corrupt transactions. The Fairfax/SBS reports helped clarify some of the mechanics of corruption in ways that could be understood by the general public. They even put together a handy graphic to explain the way politicians laundered their corrupt funds into Australia with the help of a range of intermediaries (see Figure 3.8).

Social media has also played a significant role in invigorating discussion about corruption in PNG, with numerous social media commentators exchanging stories about alleged corruption – often involving political elites. The accusations hurled around the Internet have clearly gotten under the skin of some politicians. In August 2016, a Cyber Crime Bill passed through the Parliament. At the time it passed, blogger and activist Martyn Namorong was concerned that the bill could be used to quash criticism of the

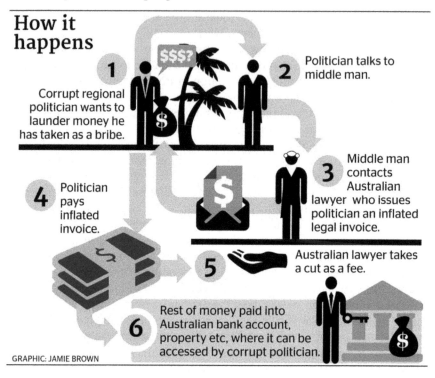

Figure 3.8 Money laundering the proceeds of political corruption
Source: McKenzie, Baker, and Garnaut (2015). Credit: Fairfax Syndication

government. The law makes reference to 'undermining the state', something he was particularly concerned about:

> You use social media to talk about protesting, for instance. Is that undermining the state or undermining the government? You know, where does it stop in a democratic society like PNG wants to be?
>
> (Radio New Zealand, 2016)

Despite these threats social media has continued to play a strong role in exposing corruption and helping to coordinate protests against it, even if much of this commentary is based on innuendo and rumour.

On the other side of the coin, there are those who argue that politicians are not as guilty of corruption as they might appear. Australian-born Dame Carol Kidu has been one of the few MPs to publicly defend politicians against accusations of corruption. Kidu relocated to Port Moresby after marrying Sir Buri Kidu, PNG's first national Chief Justice, and after entering politics in 1997 was the only female MP in PNG between 2002 and 2012. Writing in PNG's daily *National* newspaper, Kidu wrote that MPs can become embroiled in corruption because:

Aspects of the bureaucratic machinery are dysfunctional [and] people who should be going to the bureaucracy instead go to the politician for water, roads, clinics, school fees and endless needs that should not be seen as MPs' functions. Some people are desperate and some are demanding, threatening and aggressive (my staff have sometimes faced weapons and even rape threats). There is a popular perception that politicians become rich. But how many ex-politicians (or serving politicians) are rich? Many are struggling to survive because their businesses were destroyed by politics or they gave away most of what they earned as a politician – not as bribes but because they feel a Melanesian obligation and compassion for the situation many people are in.

> (From the *National* newspaper, 7 March, 2006,
> cited in Larmour, 2006: 15)

Here Kidu suggests that corruption can be a way of redistributing resources among constituents' communities. Her analysis speaks to the political dance that occurs between politicians and their constituents in PNG, which can involve candidates, politicians and communities.

The ambivalent reality of 'political corruption' is demonstrated by the controversial rise of MP-controlled Service Improvement Program (SIP) funds. SIP funds have been, at various times, allocated to local-level, district and provincial governments. The longest-running SIP fund has been controlled by PNG's 89 (out of a total of 111 MPs) district MPs. Known as the District SIP (DSIP), these funds have grown exponentially since they were introduced in the mid-1990s. In 2007 the National Government allocated 10 million kina (US$1.2 million) to each district MP in its Additional Supplementary Budget. This was augmented by an additional 4 million kina in 2009, which meant that each MP had 14 million kina to spend between the time they were elected in 2007 and the following election in 2012 (Auditor-General's Office of Papua New Guinea, 2014). In total over this period district MPs controlled 1.246 billion kina. Since the 2012 election these allocations have only increased. The newly elected O'Neill government allocated a further 10 million kina for each district MP in both the 2013 and 2014 budgets (Howes et al., 2014: 128). These funds increased to 15 million kina in the 2015 budget, giving district MPs control over a total of 3.2 billion kina between 2013 and 2015.

Civil society organisations and political commentators have been aghast at this rise in spending. Indeed, when a bill calling for an increase of con-stituency funds was raised by two MPs in 2005, civil society organisations such as TI PNG helped to scupper the draft bill by getting tens of thousands of people to sign a petition against it. But ultimately it was for naught; on the back of expanding government revenues these so-called 'slush funds' have come back bigger than ever. The naysayers are right to be concerned. DSIP funds have been found to be poorly managed. A 2014 report by the country's Auditor-General found that DSIP funds offered 'limited value' and that there had been a pervasive breakdown in governing the funds, leading to increased risk of fraud (Auditor-General's Office of Papua New Guinea, 2014).

DSIP funds have targets for how funding should be spent; the 2007 allocation of 10 million kina was nominally supposed to be spent on education, health, law and justice, water and sanitation, agriculture, rural communication and electrification (all allocated 1 million kina each, with allocations for both transport and community infrastructure totalling 4 million kina). But with poor monitoring and record keeping there is little indication that funds were actually spent in these areas. In turn, there has been some evidence that funds have been misappropriated. In 2015 a Cabinet Minister was convicted by the Leadership Tribunal for misconduct in office and misapplication of district funds – including DSIP grants. The offences, committed between 2008 and 2010, involved misusing over 230,000 kina, which were spent on office and housing rentals and a debt for car services.

Politics and culture

While flawed, SIP funding schemes are not simply a grab for money from greedy politicians. They are also a way of integrating traditional governance structures into modern-day politics. Traditionally PNG's small-scale societies were led by big-men – who gained prestige through personal attributes (such as oral prowess) and the distribution of pigs and agricultural produce – and chiefs whose position was hereditarily passed down. In both these systems leaders distributed largesse to maintain relationships; however, it is the big-man system where this is most important to gaining and maintaining status (Ketan, 2004). This has given rise to the term 'big-man politics' to describe political manoeuvring that reflects traditional political dynamics.

Ketan (2004) suggests that the Highlands big-man system has significant symbolic and functional parallels with modern state politics. His comparison of practices, symbols and spaces between 'traditional' Mount Hagen society and the modern PNG state is illuminating. He writes that like in Parliament, in Hagen, 'Important clan issues are discussed by clan members in the men's house where big-men usually dominate such meetings' (Ketan, 2004: 230). This, he suggests, mirrors the way the National Parliament works. Ketan also shows how the Melpa people of Mount Hagen gained status through what they call *moka*: reciprocal exchanges of wealth between partners. Prestige from *moka* comes from giving more than one received, which builds both group and individual prestige. This form of reciprocity meshes with the way MPs engage in pork-barrel politics, distributing largesse to shore up political support. The way in which traditional big-men in Hagen make extravagant claims of group success (*el-ik*) is similar to the way the Prime Minister and other ministers defend their actions during crises and take credit for positive outcomes. The blurring of traditional and modern forms of exchange and leadership led to the Melpa people capturing and modifying the state system to suit their life-style through, for instance, bloc-voting whereby entire clans or villages decide to vote for one candidate. This system of bloc-voting is still practised in many communities across PNG today.

This analysis highlights the difficulties of framing exchanges between politicians and their supporters as 'corruption' even if these transactions are illegal. A sympathetic reading of vote-buying, for instance, might argue that these practices are simply reflections of traditional reciprocity. In turn, the rise of SIP funding, it can be argued, is simply a way of adapting the state system to local expectations born of the norms of traditional governance systems. The highly personalised approach of MPs fits traditional governance systems far more than an impersonal approach reflecting an imagined Weberian bureaucracy (which, reflecting the mainstream view's public office perspective, considers public office as featuring a clear delineation between the public and private spheres).

Drawing on Ketan's insights, it could be also argued that the way SIP funding is used to secure political support within Parliament replicates traditional politics in some parts of the country. Since the introduction of SIP, opposition MPs have complained that the scheme has been used as a political weapon, with accusations that the ruling government has denied opposition members access to, or delayed payment of, these funds. Given that SIP funds can play a significant role in helping MPs get re-elected, in 2015 there were fewer than 10 out of 111 MPs in the opposition. This, some believe, helps secure MPs' control of this resource. Melpa big-men similarly controlled the distribution of sacred resources through selective *moka* exchanges to secure alliances between powerful individuals and groups. So, these SIP 'slush funds' can be understood as mirroring some customary forms of distribution.

The bureaucracy is similarly affected by cultural and materialist issues. Payani (2000) has argued that appointments and promotions in the bureaucracy have been shaped by *wantokism*, rather than meritocratic principles. This has resulted in what he calls an 'ethnicization of the bureaucracy', where departments are overrepresented by tribal, provincial and regional groupings. Drawing from personal observations in four government departments he notes that the Papua Region in particular is overrepresented. Payani notes that those in high positions in the public service come to represent themselves as big-men to their *wantoks*, with demands placing leaders under pressure to distribute government largesse. In this way cultural pressure, born in part from the realities of those who have missed out on the benefits of development, pressures public servants to distribute state resources.

These materialist and cultural explanations of 'corruption' are critical in a country like PNG. Yet, there is a point, some invisible line, where these explanations fail to adequately capture the reasons for abuses of power conducted by political elites. Many political and bureaucratic leaders in PNG are now highly educated, have been exposed to Western countries, and declare their commitment to upholding the rule of law and defending principles of good governance. So while materialist and cultural explanations of 'corruption' might help explain corruption between politicians and their supporters, they should have much less credence when explaining abuses of power in cooperation with other elites. When corruption involves collusion between political elites, lawyers, government departments and large business it is more difficult to postulate that those involved are primarily driven by a concern for the welfare of citizens or their cultural obligations.

Cultural and materialist explanations (as reflected in the alternative view's moral perspective) also fall flat when we consider the efforts of political elites to attack anti-corruption agencies. The method and impact of attacks on the country's Ombudsman Commission (OC PNG) are further examined in chapters 4 and 6. For now, the influence of political elites on anti-corruption efforts can be best exemplified by the case of Investigation Taskforce Sweep (ITFS). In August of 2011, ITFS was set up by newly minted PM, Peter O'Neill, soon after he rose to power to help legitimise his controversial removal of the elected PM, Michael Somare. ITFS' investigations led to the recovery of millions of dollars of stolen funds and the arrest of numerous public servants and even politicians, with three politicians receiving significant jail sentences. In 2013, ITFS moved on O'Neill over accusations that he had signed away 72 million kina (US$23 million) of the state's largesse in the form of fraudulent legal fees to a firm headed by Paul Paraka, who was himself facing criminal charges from ITFS investigations. An arrest warrant was issued, but O'Neill has so far managed to stay in power by installing a Police Commissioner sympathetic to his cause. The new Commissioner did not pursue the arrest warrant. At the same time he defunded ITFS, severely weakening its ability to operate.

This case shows how realpolitik (explored further in chapter 6), more than cultural or material factors, plays a critical role in determining the state's responses to corruption. In this case, ITFS was set up on the back of concerns about the PM's political legitimacy and then disbanded when it threatened the PM. These manoeuvres had little to do with material or cultural issues. This then highlights the importance of examining discourses of corruption at different scales. While culture has significantly shaped the state and complicated narratives about corruption, large-scale corruption between elite actors might be best explained through the mainstream view's legal or public office perspectives. How this is reflected in PNG is explored further in the findings chapters of this book. In summary, this section has focused on the importance of state actors and has found that both the alternative and mainstream views help shape narratives about and practices of corruption. However, corruption often involves actors who pay little heed to socially constructed political boundaries. As described in the following section, this is especially the case in PNG, with corruption involving transnational actors, and corrupt monies moving across state boundaries.

Transnational corruption

PNG's natural resources have long attracted transnational companies, which have exploited these resources, sometimes by engaging in corruption. This is particularly the case with the logging sector (Walton, 2013b).[5] The country has had a long history of illegal logging, which has been facilitated by an unholy alliance between logging companies and politicians. In the late 1980s the Barnett Inquiry was conducted into PNG's logging industry. It found extensive examples of human rights abuses and corruption amidst wide-scale illegal logging (Barnett, 1989). Unfortunately little has changed. One of the key organisations identified in the inquiry, Rimbunan Hijau, is still accused of bribing politicians and trampling on the rights of landowners (Laurance et al., 2011; Broadhurst, Lauchs, & Lohrisch, 2012). Rimbunan Hijau is a Malaysia-based logging company, which has investments across the Asia-Pacific. In PNG it has interests in retail and media industries, which divert attention away from its controversial logging activities. Some have traced the illegal logging concessions to the highest echelons of politics. The Barnett Inquiry linked the nation's first PM, Somare, to corruption in the logging industry, which resulted in his referral to the OC PNG for allegedly lying under oath about logging concession holdings (Chin, 2008).

The scale of illegal logging is large. Although accurate assessments are difficult, the United Nations has estimated that the net exports of illegal wood-based products from both PNG and Solomon Islands in 2010 totalled US$800 million (UNODC, 2013: vii). More recently, a US-based think tank, the Oakland Institute, has estimated that transfer pricing practices by logging companies could cost PNG US$100 million per year (Oakland Institute, 2015). PNG logs become a part of the global timber industry, as they are logged from the country for processing in Asia – to be precise, China – with approximately 90 per cent of PNG's timber exports ending up in PNG's northern neighbour (Global Witness, 2015). Once they are mixed with legal logs and made licit for consumers, illegal timber is washed of its illicit origins.

Recent government policies have exacerbated the potential for corruption in this sector through a national land-leasing scheme known as Special Agricultural Business Leases (SABL). Between 2002 and 2011 the PNG government distributed over 50,000km^2 of land under SABL agreements. SABLs were aimed at increasing agricultural production, by allowing landowners to lease their lands to businesses who would turn their land into productive farms. However, the scheme attracted loggers who came with promises of development, left with trucks full of logs and were often not seen again. Few agricultural seeds were planted. As a result Global Witness estimates that a third of PNG's timber exports in 2014 came from logging from SABL areas (Global Witness, 2015). At the time of writing there has been little evidence to suggest that logging operations in these areas have stopped or slowed. Some suggest that the inverse is true – that log exports under SABLs have in fact increased dramatically (Global Witness, 2015).

While Asian companies such as Rimbunan Hijau are most strongly associated with corruption and other nefarious activities in the logging sector, investigations into the SABL scheme have highlighted the broad array of actors involved in these activities. A Commission of Inquiry found that a number of Western companies were implicated, requiring the licenses they were issued to be revoked. This included the Australian company Leighton (PNG) Limited and Independent Timbers and Stevedoring Limited (IT&S), a 'Queensland-led' company (Garrett, 2014) described by the commission as being registered in the United States (Commission of Inquiry into the Special Agriculture & Business Leases, n.a.). IT&S garnered media attention for the size of its leases, with the company involved with four SABLs totaling more than 2 million hectares of land belonging to tens of thousands of Papua New Guineans living in the country's Western Province. The company was found to have manipulated the lease approval process and, during the inquiry, to have prepared 'deceptive and clearly fraudulent' documents (Garrett, 2014).

The international NGO Oxfam accused two Australian banks, Westpac and ANZ, of helping to finance companies involved in the SABL scheme. It accused Westpac of funding, since 1995, Malaysian logging company WTK Realty and a subsidiary, Vanimo Forest Products Ltd. The PNG Commission of Inquiry found Vanimo Forest Products Ltd had been operating in West Sepik Province without the consent of landowners (Chochrane, 2014). Oxfam has also accused ANZ of providing IT&S with an AU$220,000 bank guarantee (Chochrane, 2014).

Even schemes to save the trees have been subject to corruption involving a melange of foreigners, government officials and often unwitting landowners. The REDD – reducing carbon emissions from deforestation and forest degradation – scheme is an international approach to reducing the amount of carbon in the atmosphere. First proposed by PNG and Costa Rica in 2005, it gained international support, and subsequently REDD+ arose as a mechanism for rewarding developing countries for reducing emissions from deforestation. The practice of REDD+ in PNG has been plagued by governance scandals. Australian businessman Kirk Roberts, who was CEO of Nupan Trading, allegedly issued fake 'forest-carbon certificates' from PNG's Climate Change Office. The deal was eventually terminated by PNG landowners in Western Province, and led to the removal of the head of the country's Office of Climate Change (Wilkinson & Cubby, 2009). The story, not the outcome, is apocryphal of the corruption that has gone on in the effort to ostensibly save PNG's forests from deforestation. Promises of future riches have lured what some in the media have dubbed 'carbon cowboys' to PNG who have colluded with PNG's mismanaged Climate Change Office. The resultant corruption has facilitated further environmental devastation.

Money laundering and the West

As explained earlier in this chapter, Australia has become a destination for corrupt funds from PNG. So much so that in 2012 Sam Koim, then new to his role as head of the country's ITFS, called Australia the 'Cayman Islands of the Pacific' for money laundering at a presentation to Australian

Transaction Reports and Analysis Centre (AUSTRAC) – Australia's anti-money laundering agency. The laundering of money from PNG to Australia has been a concern within law enforcement and regulatory agencies for some time. In 2008, AUSTRAC noted that, 'Australia is a significant destination country for funds derived from corrupt activities within the region' (AUS-TRAC, 2008: 3). As Sam Koim said a few years later, the corrupt in PNG see Australia as the 'preferred place to launder and house the proceeds of their crimes because it is … an extremely easy process to get their money into the Australian system' (Koim, 2013: 246).

An example of money laundering from PNG to Australia is the case of the 2009 sale of 500 million shares from PNG-owned Motor Vehicles Insurance Limited (MVIL), managed by an Australian investment company, Woodlawn Capital. Not only did the shares managed by Woodlawn drop in value by AU$10 million, company managers demanded AU$23million in fees (Callinan, 2014). The sale was deemed illegal under PNG's Independent Public Business Corporation Act, yet was approved by the then Minister of Public Enterprises – and son of the then PM – Arthur Somare (Sharman, 2012: 9). It has been alleged that Woodlawn may have been controlled by the powerful Somare family (Sharman, 2012). Given the illegal nature of the payment and the links to powerful elites in the country, it is alarming that the Australian anti-money laundering system failed to detect this payment, with AUSTRAC investigations only occurring *after* the PNG government, under PM O'Neill, broke the story in late 2011 (Sharman, 2012: 9). The PNG government brought the case to the New South Wales (Australia) Supreme Court, which in 2015 ordered the company to pay back AU$20 million to the MVIL (Lapauve, 2015). This example again highlights that corruption in PNG must be understood as involving an array of actors, including those from the West (and Australia in particular), who help facilitate corruption and hide the proceeds of these nefarious activities.

Porous borders

The involvement of foreigners and transnational syndicates in corruption has greatly increased given lax controls and high levels of corruption within PNG's immigration and customs departments. Passport scams in PNG have been a long-running problem. In 2003, it was revealed that a machine used to print passports was stolen during a break-in at the Foreign Affairs Department. In the same year, a report into the Immigration Department found that officials, in conjunction with foreign nationals, were improperly issuing passports and visas to allow illegal immigrants into the country or to help them find passage to third countries, like Australia (McLeod, 2003). In 2009, then PM Somare admitted that people could get a passport through the Immigration Division by bribing officials with a six pack of beer. He said that, even as PM, he had had to wait four or five days for his passport, while others saw much quicker turn-around times.

Somare intimated that this discrepancy was because others were willing to pay officials to speed up the process (*PNG Post-Courier*, 2009).

A few years later, in 2012, one of the country's daily newspapers, the *PNG Post-Courier*, highlighted the dismal state of border control. Immigration and Foreign Affairs officials admitted that the entry of illegal immigrants had gotten out of hand as their office had been 'compromised' and had few resources. The paper quoted an anonymous source as saying:

> We have lost control of the issue of immigrants coming into the country as early as 10 years ago. It's not because we cannot handle it, it's because we do not have the resources and finance to tackle the issue very effectively.
>
> (*PNG Post-Courier*, 2012a)

According to documents obtained by the paper around 11,000 'Asians' entered the country every 5 years and 30,000 were 'floating', squatting' and working illegally throughout PNG. Citing documents obtained, the paper reported that, as a result, three Asian gangs had established businesses across numerous industries including real estate, hospitality and security companies. While the numbers of illegal immigrants and crime gangs are to be taken with a few grains of salt, such constant reporting of dysfunctional border control has led to broad-based disquiet about new Asian immigrants. Some in the Asian community are seen by Papua New Guineans as both a product of and contributor to corruption. Chapter 4 shows how concerns about these illegal entries, and corruption associated with the newer members of the Asian community, resulted in riots across the country which were facilitated by one group of Papua New Guinean anti-corruption activists.

Also concerning are the links that are purported to exist between illegal immigrants and politicians. The *PNG Post-Courier* exposé noted that some of those entering into the country illegally were working in partnership with elite leaders and MPs. The case of fugitive businessman Joko Tjandra highlights the problem of political interference around immigration policy. Tjandra, an international fugitive, who is wanted by authorities in Indonesia for tax fraud, is listed by Interpol for corruption-related charges (Interpol, 2009). In 2009 he fled Indonesia before his sentencing over AU$57million payment he received from the sale of Bank Bali; he was convicted and sentenced to two years *in absentia* (Callick, 2012). Despite his background he has been feted by PNG MPs, with government ministers lured in by Tjandra's promises and extensive business holdings (Minta, 2014). He was flown from Malaysia to PNG in 2011 and granted PNG citizenship, despite serious concerns raised within PNG about his eligibility (Minta, 2014). Tjandra's company, Naima Agroindustry, has since been awarded an effective monopoly over the country's rice market. However, with the country experiencing severe drought in 2015 and thus needing all of the food it could get, at the time of writing it is still not clear if the project – and monopoly – will go ahead. This example highlights the ability of foreign interests to evade immigration standards and an extradition treaty, and to capture the state.

Poor immigration and customs services mean that it is not just people that cross PNG's borders illegally. Contraband of various sorts crosses between PNG and its neighbours. Across the Torres Strait sea border with Australia, drugs, weapons and other illicit goods have been smuggled. The land border between PNG and Indonesia is also regularly breached, with contraband moving between the two countries freely. The PNG–Solomon Islands sea border is similarly porous. The existence of these illegal trade routes has often been associated with corrupt officials who are willing to look the other way. In 2006 the Australian Broadcasting Service's (ABC) *Lateline* programme investigated the trade of guns, drugs and people through the Torres Strait. Those involved in smuggling accused Australian officials of facilitating smuggling. Australian immigration, customs and police officials were accused of dealing drugs and taking bribes (Bohane, 2006).

Transnational corruption therefore involves numerous actors, all of whom are differently enfranchised. Some foreign actors are feted by the government and given protection from powerful backers. Others engage in small-scale crimes, or enter the country illegally with few resources. There is a vast difference between genuine refugees using PNG as a launching pad to arrive in Australia, and the nefarious dealings of business elites. Unpacking the nature of transnational corruption and related crime then reveals the embedded power relations which frame corrupt activity. In turn, it shows that not all corruption is equal.

Conclusion

This chapter has shown that corruption in PNG has a scalar dimension, in that it involves actors who operate at different social, administrative and regulatory scales. At each of these scales corruption involves different amounts of money and resources – corruption ranges from petty, through to grand and systemic types. Examining these scales and connections between them matters for understanding the applicability of the alternative and mainstream views (and their associated perspectives) presented in the previous chapter. At the local level we see that the alternative view is critical for understanding marginal concerns about corruption. Given the weakness of the PNG state, responses to corruption are often – though by no means always – determined by cultural considerations and economic marginalisation, which reflect the moral perspective. At the local level it is also important to understand the power dynamics that can shape people's willingness to respond to corruption. Still, the findings reveal that even at the local level the way many citizens frame corruption accords to the public office and legal perspectives of the mainstream view. This is particularly the case for more enfranchised respondents.

When we lift our gaze to the national scale, the tensions between the mainstream and alternative views on corruption are played out through national politics. The alternative view's concern with culture is apparent in the justifications for the country's massive constituency (SIP) funds. These funds are justified in the name of 'big-man' culture, but allow MPs to solidify their patronage networks and have resulted in the centralisation of power among government MPs. The rise of constituency funds in PNG then, cannot be understood apart from the socio-cultural tropes which justify them. By replicating local forms of governance they have successfully gained local and national support, and thus have helped bridge the gap between local needs and national resources, albeit in highly uneven ways. They have in effect legalised a Papua New Guinean moral perspective by providing resources to political big-men. PNG's laws themselves have bent to the cultural concerns of the alternative view.

The mainstream view suggests that (from the economic perspective) the opening up of the economy should help reduce corruption. However, the nature of transnational corruption involving PNG highlights the limitations of this view. Economic globalisation has increased the chances for corruption between a nexus of private and state actors (Walton & Dinnen, 2016). These groups often operate behind a veil of legitimacy, which means that establishing new laws or trying to enforce existing laws can be a fruitless exercise. Appeals for elites to reflect the public office perspective by being loyal to state norms are also unlikely to be heeded. In this context there is thus a need to draw upon the critical perspective to better understand and respond to corruption that facilitates the transnational circulation of illicit money and goods. Examining the ways power is expressed at the international level, then, is critical for understanding why elites engage in corruption so freely. Responding to elite corruption at this scale will require addressing the nexus of political power that is bound up within these transactions. This will mean examining and responding to the political power of a range of state and private sector elites who move between nation-states.

In sum, addressing corruption in PNG will require anti-corruption actors to understand and respond to both the mainstream and alternative views on corruption at a variety of scales. While the mainstream view is popular in international anti-corruption circles, alone it does not offer an adequate framework in PNG due to the country's history, poverty and diverse cultures. The next two chapters examine the ways international and national efforts have responded to the challenges of corruption in PNG. They take particular note of the way anti-corruption organisations and actors reflect the mainstream and alternative views through their discourse and activities.

Notes

1 While the amount of money involved in this transaction is not indicated, it is assumed that the amount would be of a large scale given the nature of the transaction.

2 As noted in the literature review, the 'decay' definition features both institutional and individual types. The Homebrew scenario alludes to one aspect of this definition: possible individual decay.

3 It is worth noting that many women in PNG still engage in the political process, such as elections, despite these poor electoral outcomes (McLeod, 2002).

4 We included eight out of the nine scenarios in our model, leaving out the Homebrew scenario, in order to reflect the types of corrupt activities which would be responded to by state authorities.

5 As I explain elsewhere (Walton, 2013b), this is not to say that other industries – such as mining – are not corrupt, but corruption is not as well-documented in these other private sector industries.

4 The nation's guardians

National anti-corruption organisations' views on corruption

Introduction

This chapter examines the activities and discourse of two key anti-corruption organisations in PNG. The first is the Ombudsman Commission of PNG (OC PNG), the country's oldest anti-corruption institution. The OC PNG is an industry insider, it is supported by the Australian aid programme and it occasionally works with TI PNG. The views of this anti-corruption stalwart are compared to those of the Coalition, a group of protestors who took their concerns about elite corruption – particularly that of Prime Minister Michael Somare – to the streets. The chapter examines the ways these organisations reflect the mainstream and alternative views on corruption.

The chapter first examines the way this young nation framed corruption and its pathway to development leading up to and in the early years of its independence. The following section introduces the OC PNG, and presents its perspectives on corruption. The Coalition and its perspectives on corruption are presented next. The chapter concludes by discussing the implications of these findings.

Corruption, development and the PNG constitution

In 1972, three years before independence, the Constitutional Planning Committee (CPC) was set up to make recommendations on a constitution for the newly independent nation of PNG. The CPC held over 100 meetings around the country, attended by 60,000 people. Thousands of submissions in verbal and written form were also received. The resultant report was released in 1974. Examining this report provides an insight into some of the aspirations and concerns of people who were about to move into a new post-colonial era.

To start with, the CPC was well aware of the threat that corruption posed to the country. Chapter 3 of their report states that corruption is a serious threat to PNG, given the experience of other nations: 'Corruption in public life is, of course, a worldwide problem which has reached very serious proportions in a substantial number of countries – developing and industrialised alike' (Constitutional Planning Committee, 1974: 3/23). What is noteworthy is how much attention the report pays to corruption in *developed* (industrial) countries. Drawing on the lessons of the Watergate scandal in the United States, the report suggests that this is evidence that 'gross abuses of political power have infected the whole American political system' (Constitutional Planning Committee, 1974: 3/24). It goes on to proclaim that public life has been subject to corruption for decades. It also highlights deep-seated local-level corruption in Britain. When developing countries are mentioned – such as Trinidad and Tobago or Zambia – it is in reference to how they proposed to address corruption.

Emphasising the problems of corruption in Western countries is, of course, very different to the official messages of the anti-corruption industry, which stresses that poor developing countries are corrupt, and rich countries are much less so and thus hold the answer to corruption in developing countries. This message is a conclusion that emanates from TI's Corruption Perceptions Index, where the poorest countries in the world – including PNG – regularly appear towards the bottom of the list. And it is the countries of the rich world who have been nominated as anti-corruption saviours; in the case of PNG it is Australia who most prominently plays that role. What is particularly interesting in the CPC's approach to describing corruption is that it sought to learn from the failures of developed nations; not just their perceived record of success. The emphasis of the CPC report can be read as an early challenge to Western-centric responses to corruption which mirrors the critical perspective.

The CPC's report is also important for debates about local responses to corruption because it clearly articulates the necessity of building the new nation through melding traditional and Western epistemologies. Chapter 2 of the report outlines 'Papua New Guinean Ways' – a set of principles that the CPC believed should guide development. Papua New Guinean Ways were framed as a response to the overwhelming experience of colonialism, which is compared to a tidal wave:

> [Colonialism] has covered our land, submerging the natural life of our people. It leaves much dirt and some useful soil, as it subsides. The time of independence is our time of freedom and liberation. We must rebuild our society, not on the scattered good soil the tidal wave of colonisation has deposited, but on the solid foundations of our ancestral land. We must take the opportunity of digging up that which has been buried. We must not be afraid to rediscover our art, our culture and our political and social organizations.
>
> (Constitutional Planning Committee, 1974: 2/98)

The report suggests that the new nation should adapt some of the best parts of the colonial experience: 'We should use the good that there is in the debris and deposits of colonisation, to improve, uplift and enhance the solid foundations of our own social, political and economic systems' (Constitutional Planning Committee, 1974: 5/99). Undesirable aspects of

Western and traditional ways and institutions should, the report argued, be left behind.

The CPC wrote clearly about the potential for state officials to fall victim to various types of corruption, including bribery and gifts to leaders, and corrupt engagement with foreign companies. The CPC's concerns about corruption and the importance of Papua New Guinean ways made their way into the structure of the OC PNG and its mandate (de Jonge, 1998). The CPC recommended that the new constitution feature a Leadership Code and an Ombudsman Commission (the OC PNG), with the OC PNG given sole responsibility for enforcing the code – which, as elaborated later in this chapter, provided a code of conduct for state officials – as well as other functions aimed at increasing transparency and the accountability of state officials. This recommendation made it into the constitution, although specific mention of corruption was not transferred across (the Constitution refers to 'wrong conduct' instead).

Not everyone was happy about the OC PNG and the Leadership Code being built into the nation's constitution. Julius Chan, who was Finance Minister at the time of independence and was later to become the nation's Prime Minister (twice), has recently stated that he was uncomfortable with this arrangement. In his autobiography he suggests that it reinforced 'a view that the elected representatives of the people could not be trusted'. He implies that the inclusion of these institutions was brought about due to the racism of 'academics and saintly bureaucrats' (Chan, 2016: 65). Yet his book is full of examples of corruption involving political elites, and the problems associated with corruption. Without the OC PNG and Leadership Code being enshrined in the constitution, it is difficult to see how these institutions could have withstood the frequent attacks from PNG politicians. Corruption would likely have been worse without them.

The Constitution also includes an ode to the CPC's concern about the nation's values, through a brief nod to the importance of development being guided by Papua New Guinean Ways. These are described as a set of values that considers:

> the cultural, commercial and ethnic diversity of our people is a positive strength, and for the fostering of a respect for, and appreciation of, traditional ways of life and culture, including language, in all their richness and variety, as well as for a willingness to apply these ways dynamically and creatively for the tasks of development.
>
> (Independent State of Papua New Guinea, 1975: 4)

This call for strength through diversity was meant to augment the Western-oriented institutions (including Parliament, the courts, police force, and the constitution itself) so central to building a modern state. The Constitution also acknowledges the importance of Christianity – another institution imported from the West. Indeed, the Constitution states that the people of

PNG will pass on both 'noble traditions and the Christian principles that are ours now'.

The emphasis that the Constitution and the CPC, in particular, placed on developing a hybridised approach to statecraft was further expanded by PNG statesman Bernard Narokobi (1937–2010). Narokobi was a permanent consultant to the CPC and served as a government minister in two governments between 1988 and 1994, and then in opposition until 2002. Narokobi spoke of the importance of 'Melanesians' – for Narokobi this grouping was made up of PNG, Fiji, Solomon Islands, New Caledonia and West Papua (Irian Jaya) – drawing on the 'Melanesian Way' as a unique form of pan-nationalism, which is comprised of a diverse political ontology combining Christian, 'traditional' and modern values and stressing self-reliance. For Narokobi, Melanesians are 'guided by a common cultural and spiritual unity' which differentiates them from 'Asians or Europeans' (Narokobi, 1980: 7). He drew on the importance of this unity to argue against imposed Western institutions – of governance, religion and economy – that degrade Melanesian ways and to call for PNG (and by implication other Melanesian countries) to develop institutions that reflect the unique needs of Melanesian citizens. Narokobi did not condone corruption. Indeed, he is credited as one of the earliest post-independence politicians to speak out against it (Joku, 2010). Yet he did argue against those who viewed customary ways as a cause of corruption (Narokobi, 1989). For Narokobi, PNG's diverse cultures could strengthen the state's ability to provide just outcomes for its people.

There has been much debate about the relevance of the Melanesian Way for the modern PNG state. For some the term is employed by politicians to help them get away with corruption and abuse the rule of law. For example, in the early 2000s Pitts summed up this view, arguing that the 'Melanesian Way' has become 'a soft term for fraud, nepotism and cronyism that denies citizens their rights' (Pitts, 2002: 177). Larmour highlighted this problem by noting that a previous Prime Minister of the country refuted criticism over payments he made 'to prevent a minister defecting to the opposition, by arguing that "gift giving" is part of the Melanesian political tradition' (Larmour, 1997: 2). This pessimism has likely dissuaded scholars from exploring the alternative view. Controversial political commentator, Susan Merrell, an outspoken critic of the Melanesian Way, has said that the concept is too nebulous to be useful, and suggests that the PNG state is failing due to its attempt to mix traditional and modern systems. The messiness of this hybrid system means that the 'big-men' of PNG politics flout PNG's law in the name of custom (Merrell, 2012). As outlined later in this chapter, these debates are still central to local responses to corruption, with both the OC PNG and the Coalition grappling with the relevance of Papua New Guinean and Melanesian Ways for fighting corruption.

The OC PNG

The OC PNG had, at the time of fieldwork, four offices around the country – in Port Moresby, National Capital District; Kokopo, East New Britain Province;

Mt Hagen, Western Highlands Province; and Lae, Morobe Province. It was headquartered in Deloitte's tower, a 15-storey high-rise, a block south from the then offices of TI PNG. The tower itself was caught up in one of the country's worst corruption scandals, making it a somewhat ironic location for the nation's foremost anti-corruption watchdog.

In the late 1990s Kumagai Gumi, a Japanese construction company, was contracted by PNG's government-run national superannuation endowment, the National Provident Fund (NPF), to build what is now known as Deloitte tower. The profitability of the project was undermined by the devaluation of the PNG kina and project overruns, leading the company to apply for a devaluation claim against the NPF. This is when an alleged fraud took place, allegedly involving the current PM, Peter O'Neill, then NPF chairman, Jimmy Maladina, and NPF lawyer Herman Leahy. A subsequent report into this and other alleged corruption within the NPF found that Kumagai general manager, Shuichi Taniguchi, paid 2.5 million kina to Maladina, O'Neill and Leahy, out of a 5.8 million kina devaluation settlement. The report found that the 2.5-million-kina payment was:

> spurious and was agreed upon between Mr Maladina and Kumagai Gumi managers (under pressure from Mr Maladina) to enable the money to be channelled through Kumagai Gumi and on paid for the benefit of Mr Maladina, with shares for Mr Leahy and Mr O'Neill (through the account of Carter Newell...). NPF is entitled to recover this K2.505 million.
>
> (National Provident Fund Final Report, 2015)

In May 2015, after fleeing the country and then eventually returning after years in hiding, Maladina was found guilty of fraudulently acquiring the 2.5 million kina – 17 years after the case was first brought to court. Charged with misappropriation, O'Neill was referred to the Waigani Committal Court in 2005, but the charges were dropped due to insufficient evidence. The National Court issued a permanent stay on the prosecution of criminal charges against Herman Leahy in 2014, finding that the Public Prosecutor had not afforded him a fair trial within a reasonable time.

Kumagai paid a 3-million-kina out-of-court settlement to NPF's board of trustees in 2006 (*PNG Post-Courier*, 2006). The company's motto is 'excitement to our customers' – the fraud it undertook in PNG certainly added excitement to the lives of those involved. Perhaps because of the NPF scandal its president, Yasushi Higuchi, now stresses that the company only engages in 'honest business practices' (Higuchi, 2015). The NPF tower scandal and others associated with the superannuation fund meant that PNG's government workers lost around half their superannuation entitlements. This led to a restructuring of the NPF, which meant it was effectively privatised and became known as the National Superannuation Fund (NASFUND), which was incorporated as a company under the *Companies Act* in May 2002

(NASFUND, n.d.). The fund is now run much more effectively than its pre-decessor, and has provided members with good yields; however, the scandal provides those who work for the OC PNG with a constant reminder of the enormous battle they face in tackling corruption.

Mandated and resourced to fail?

The OC PNG grew out of recommendations from the Final Report of the CPC (Constitutional Planning Committee, 1974), with chapter 11 outlining the role the OC PNG would play in a newly independent PNG. The CPC's Leadership Code recommended rules aimed at holding Papua New Guinean leaders (defined broadly) to a morally and legally binding code of conduct; it recom-mended that the OC PNG be responsible for overseeing the code and seeing that leaders comply with its provisions. A version of the Leadership Code was adop-ted in the 1975 PNG Constitution (de Jonge, 1998), with the OC PNG given sole responsibility for enforcing it as well as other functions aimed at increasing transparency and the accountability of state officials.

Since independence the OC PNG has tackled corruption and mismanage-ment, including constitutional breaches by the National Executive, the National Parliament, constitutional office holders, PMs (including past PM Somare), and other MPs and members of the judiciary (Justice Advisory Group, 2005: 4). However, for an organisation that promotes transparency the OC PNG struggles to live up to its own aspirations. At the time of writing the organisation had yet to publicly release eleven years of annual reports (from 2006 to 2016) – although it does communicate to the public through the media and a glossy newsletter. We do know that, in 2005, the OC PNG received and dealt with 3,299 complaints through its four offices, with over half dealt with by its head office in Port Moresby. The head office in Port Moresby was involved in 1,442 complaints; the regional offices in Mt Hagen, Lae and Kokopo reported 610, 682 and 565 complaints, respectively (Ombudsman Commission of Papua New Guinea, 2006: 35). In 2005 the OC PNG had 125 fully funded staff positions, although the organisation only filled between 101 and 103 of those positions during the year (Ombudsman Commission of Papua New Guinea, 2006: 20).

Complaints varied in size and complexity. The OC PNG has conducted in-depth investigations into large-scale fraud such as the purchase of the Cairns Conservatory by the Public Officers Superannuation Fund Board at a grossly inflated price (Ombudsman Commission of Papua New Guinea, 1999), through to relatively small complaints, such as the parking conditions on Douglas Street in downtown Port Moresby (Ombudsman Commission of Papua New Guinea, 2006: 41). But it is political cases that gain the most attention – the ways the OC PNG has responded to these cases is detailed in chapter 6, but for now suffice it to say that the agency has struggled to investigate allegedly corrupt politicians without suffering from the fallout. This fallout can be particularly damaging given that the PNG government is the OC PNG's main funder.

Despite a tumultuous relationship with political leaders, limited powers of censure and meagre resources (Justice Advisory Group, 2005), by 2008–09 the OC PNG had become one of the only stalwarts against corruption in the country. The organisation complements other state-based institutions such as the Public Service Commission, the Judicial Service Commission, the Electoral Commission and the Auditor-General's Office. However, its ability to keep state officials accountable was severely strained by lack of resources. This became particularly acute after the 1995 reforms to the Organic Law on Provincial and Local Level Governments. For the OC PNG this Organic Law meant that the organisation was now responsible for investigating members of provincial assemblies and local-level governments, who had up until that point not been subject to the Leadership Code. The new Organic Law – a foundational law that is secondary only to the constitution – led to the creation of almost 300 new local-level governments, and an additional 5,340 politicians (de Jonge, 1998). While the OC PNG was responsible for keeping a check on these leaders it did not receive a corresponding increase in resources (de Jonge, 1998). The OC PNG has also come under attack from politicians eager to curb its powers – the way civil society actors protested against a bill to reduce the power of the OC PNG is further discussed in chapter 6.

The OC PNG is headed by three Ombudsmen – one Chief Ombudsman and two ordinary Ombudsmen – often working against great odds to investigate corruption. When I was conducting fieldwork Chronox Manek was the Chief Ombudsman Commissioner, with the late Ombudsman John Nero and Ombudsman Tabitha Suwae by his side. Chronox Manek was shot in the arm in 2009 – in what many suspected was an attack by those he was investigating – and after suffering from poor health died in 2012. This left a vacuum within the commission, until Rigo Lua was appointed in mid-2013; Lua reinvigorated the OC PNG, by leading it on investigations into prominent politicians and public servants. However, in May 2015, it was revealed that Lua's appointment would not be renewed by a committee including prominent politicians. It is likely that in the eyes of many of PNG's elite Lua was considered too successful.

The OC PNG, in other words, had a weak mandate, was subject to great political pressure, and was struggling to live up to the levels of transparency and accountability expected of a modern ombudsman commission. Despite these limitations, the OC PNG continued to operate and tried to keep state officials accountable insofar as it could.

Official definitions, causes and responses

While the CPC was explicit in its concern about corruption, the Constitution of PNG does not mention the word directly. It does however use the term 'wrong conduct', a term the OC PNG equates to corruption. Section 219 (2) of the constitution states that 'wrong conduct' (corruption) is:

a   contrary to law;
b   unreasonable, unjust, oppressive or improperly discriminatory, whether or not it is in accordance with law or practice;
c   based wholly or partly on improper motives, irrelevant grounds or irrelevant considerations;
d   based wholly or partly on a mistake of law;
e   conduct for which reasons should be given but were not, whether or not the act was supposed to be done in the exercise of deliberate judgement within the meaning of Section 62 (decisions in *'deliberate judgement'*).

This interpretation of 'wrong conduct' sets out both legal and immoral activities that fall within the OC PNG's remit: acting 'contrary to law' (point a) is a concern of the mainstream view's legal perspective, while 'improper motives' (point c) relates to the moral perspective of the alternative view. Here 'improper motives' include conflicts of interest, favouritism, bad faith and dishonesty (Ombudsman Commission of Papua New Guinea, 2001: 25).

What makes the OC PNG unique in relation to other ombudsmen commissions around the world is this focus on the *moral* conduct of state officials. This means that politicians, for example, can be and have been referred to the Leadership Tribunal for acts that are immoral, even if they are not illegal. For instance, in July 2015 MP Delilah Gore was found guilty by the Leadership Tribunal for failing to switch off her mobile phone, which is contrary to the PNG Civil Aviation Rules of the national airlines operating procedures. Critical to the Tribunal's judgement was that her disorderly conduct – specifically Gore's argument with the flight attendant – brought her position as leader into disrepute. Here we see that it was not just the breaking of laws and rules that was important to the decision, it was also the way that the MP conducted herself that led to a guilty verdict.

The OC PNG's definition of corruption is tightly tied to public officials. Section 13 of the Organic Law of the Ombudsman Commission states that the OC PNG's jurisdiction extends to investigation of:

a   Any state service or member of any state service; or
b   Any government body or an officer or employee of a governmental body; or
c   Any other service or body referred to in Section 219 (a) (Functions of the Commission), that the Head of State, acting with, and in accordance with the advice of the NEC [national executive council], by notice in the National Gazette, declares to be a service or body for the purposes of this section. (Ombudsman Commission of Papua New Guinea, 2001: 9)

When proposing the Leadership Code, the CPC noted that 'leaders' in the public sector were the focus of their concerns rather than those in the private sector because 'persons who hold office in private concerns ... are not the servants of the people of the nation' (Constitutional Planning Committee, 1974: 3/1). The OC PNG is concerned with 'decisions or actions contrary to

Organic Law, Act or Regulation [... and the] failure to comply with common law; incorrect application of the law; and incorrect interpretation of the law' (Ombudsman Commission of Papua New Guinea, 2001: 22). The organisation has also highlighted that corruption can be caused by decisions or actions 'inconsistent with adopted guidelines or policy' of a state agency (Ombudsman Commission of Papua New Guinea, 2001: 23). In turn, the OC PNG essentially defines corruption as the equivalent of 'the abuse of public office for private gain', and is concerned with its legal and bureaucratic causes, reflecting the public office perspective (see chapter 2), albeit with a moral twist.

As described in chapter 2, the public office perspective of the mainstream view prioritises responding to corruption that involves public officials transgressing norms, rules and laws. The OC PNG follows this logic as a primary consideration for evaluating the need to respond to corruption, even though it acknowledges that corruption is also a moral issue. For example, if, after an investigation of a leader, the Leadership Division finds prima facie evidence of misconduct in office, the OC PNG is required to refer the matter to the Public Prosecutor (which decides on whether to prosecute a matter though the Leadership Tribunal) or directly to the appropriate tribunal for prosecution (Ombudsman Commission of Papua New Guinea, 2001: 77). These tribunals determine whether a leader is guilty of misconduct in office and, if guilty, recommends punishments that mostly involve sanctions related to an individual's post. These may include dismissal from office, monetary penalty, suspension from duties or deduction of salary (Ombudsman Commission of Papua New Guinea, 2006). So, a decision to investigate is determined by the relevant laws that govern a leader's position, and the punishments relate to curbing the entitlements provided by the state.

In an undated report, John Wood presented a brief assessment on the achievements of some of the Ombudsmen across the Pacific. He found that out of the 93 leaders the OC PNG helped refer to the Leadership Tribunal between 1978 and 2010, only 4 were found not guilty (while 8 were pending). A third of leaders quit politics or lost their seat before the tribunal was appointed or completed its hearings – this allowed them to dodge punishment. That left 26 dismissals, 14 with fines, 1 suspension and a reprimand (Wood, n.d.: 16). For those who were penalised, the punishment failed to have much of an impact. Wood's analysis is complemented by Ainsley Jones (2014), who examined the political fallout from a number of cases brought before the Leadership Tribunal. By tracking the career of politicians who had been found guilty by the Tribunal, she was able to get a sense as to whether or not guilty politicians' careers suffered. For the most part Jones found that these MPs suffered very little political fallout. Indeed, some went on to hold very senior portfolios within the government. While acknowledging that there may be some shame suffered by MPs from the resultant media exposure, over the long term, being found guilty did not concern their constituents enough to not vote for them. Nor did it sufficiently undermine the faith their colleagues had in their ability to manage key ministries.

As discussed in chapter 3, citizens are far more likely to report corruption in PNG if they believe that the state is able and willing to address it. The OC PNG has oftentimes failed to assure citizens that it is capable of providing a substantive response to corruption, despite its continued and sometimes brave attempts to do so (which are further explored in chapter 6). In turn, it is likely that its failures dampen the willingness of citizens to engage in anti-corruption activities. Given the OC PNG's inability to keep state officials accountable, the dreams of the CPC and PNG's founding leaders have by and large been unrealised.

Underneath the veneer

While the OC PNG has failed to live up to the hopes of the nation's founding leaders, there are still many who believe that fighting corruption should involve drawing on the strength of PNG's many and varied cultures and traditions. This was the case in interviews with key stakeholders within the OC PNG, many of whom were convinced that PNG values, enshrined in the Papua New Guinean Ways and Melanesian Way, could play a role in framing responses to corruption in the country. This was evident from the way in which respondents defined corruption, with most reflecting the moral perspective. A senior employee who was familiar with investigating the legality of corruption cases stated that, 'my simple definition of corruption is doing things out of the norm'. 'Doing things out of the norm' is different to acting illegally or not following due process, it suggests that for this respondent corruption must be understood through reference to societal, rather than state-based, institutions. Similarly a senior respondent suggested that 'people have their own views of what corruption is', which, he said, made it hard to know the types of acts it did, and did not, include.

While all respondents were concerned that the state's top politicians and bureaucrats were responsible for corruption, many believed that, ultimately, their alleged crimes were caused by broader structural issues, which in effect mitigated their individual responsibility. For many, politicians and bureaucrats were seen as corrupt because they did not know what was expected of them. An expatriate advisor said:

> The basic administrative expectations that [Australians] have ... that public servants will know the law that they work under, and work accordingly, does not happen ... and it is the biggest problem with corruption here ... We are finding agencies where there are no records, where no-one knows what anyone should be doing and there is no way of tracing whether the person that made a decision had the authority to make that decision.

This respondent went as far as insisting that the OC PNG did not actually investigate corruption, because 'corruption' suggested pernicious selfishness. In his view, a bureaucrat could not be acting out of self-interest if they didn't know that they were doing the wrong thing. Other respondents suggested that MPs were similarly confused about their roles. For instance, a senior employee said that because few MPs understood the formal laws governing their position, it was natural for them to revert to their familial obligations and 'distribute as much as they can' to their supporters, whether this is legal or not.

Respondents were also concerned that public servants' poor pay exacerbated the potential for corruption. One respondent from one of the OC PNG's regional offices suggested that corruption was caused by 'not enough pay or remuneration ... to keep public servants going'. According to this respondent, not getting paid enough was a problem because there was not enough money to take care of demands by friends, family and *wantoks*. Explaining the connection between low pay and corruption, another respondent said if a public sector employee could not meet 'certain social obligations ... he needs to put his hand into the public purse'. For these respondents corruption was caused by the inability of individuals to meet social and cultural obligations.

Again reflecting an alternative view on corruption, respondents complained about the influence of Westernisation as a driving force for corruption. A respondent from one of the OC PNG's regional offices said corruption was caused by 'Western ideas or conceptions'. He suggested that corruption occurred because Western materialism causes Papua New Guineans to start thinking, 'I want to be like that bugger with the high class car or living in a good house.' A senior employee from Port Moresby suggested that corruption was relatively new to PNG and was introduced along with a range of other issues that were not traditionally a part of Papua New Guinean life: 'a lot of new things are coming in from the outside: the new concepts, the people have been educated; there's a whole range of developments taking place and corruption is one of those things.' In the eyes of many associated with the OC PNG, Westernisation has meant that, as another senior employee put it, 'Papua New Guinean culture has become corrupted.' For some this resulted in ambivalence towards the relationship between culture, development and corruption. A senior respondent with decades of experience in public service said:

> In many, many instances culture has acted as a buffer to maintain some level of decency ... it is what being a human being is about. We'd like to protect [these aspects of culture] even if we have to do it with our lives. On the other hand, [some aspects of] culture have been so corrupted almost out of sight ... people seek refuge in corruption to explain that [corruption] is done in the culture, and that sometimes ... cuts across laws that are brought in to prevent [corruption].

In other words, in the minds of these respondents, PNG's culture was something to be protected and valued. When cultural obligations were considered a part of the problem, they were tied to poor remuneration rather than the practice of looking after kith and kin. Yet while cultural obligation, poor remuneration and Westernisation ameliorated the responsibility and agency of those engaging in corruption, these factors did not erase concerns about corruption. Respondents did label those engaging in corruption as ultimately 'selfish', 'greedy', 'self-centred' and 'deceitful'. But they longed for a more nuanced understanding about the causes of corruption that accounts for the structural pressures facing those in government office. These concerns were noticeably absent in the OC PNG's official documentation which mostly – though not exclusively – focused on the agency's legal mandate and the legal responsibilities of state officials.

In turn, there was some discussion about the need to look to PNG's diverse cultures to frame responses to corruption, which was evident in a speech given by the then Ombudsman Commissioner, Ila Geno, during the launch of an edited book on corruption. In March 2008 Geno was asked to speak at the launch of the National Research Institute's collection of articles on corruption: *Corruption in Papua New Guinea: Towards an Understanding of the Issues* (Ayius & May, 2008). The Chief Ombudsman told the assembled crowd that as a boy growing up in the village he took a coconut from a tree and brought it back to his father. His father recognised the coconut as belonging to another villager, and chastised the young Geno for taking it. The story's moral, Geno stressed, was that all Papua New Guinean cultures had developed understandings of right and wrong, thus all Papua New Guineans could understand corruption. The similarities between otherwise diverse Papua New Guinean cultures could be drawn upon to fight corruption and make the nation stronger. Geno's narrative was, in effect, a call for a positive form of nationalism: one where citizens unite through their shared moral understandings. It recalls Narakobi's Melanesian Way, or the Papua New Guinean Ways of the constitution.

Culture then was central to any discussion about how to address corruption in PNG. It was evoked by those within the OC PNG to justify the organisation's legitimacy and to guide and inspire anti-corruption efforts. Some also stressed the importance of accounting for the aspects of modern life that were bastardising culture. For some, reducing corruption meant addressing the material conditions facing citizens and officials so they could afford to say no to corruption. A senior employee suggested:

> To lessen the instance of corruption on a large scale, we need to look at socio-economic aspects and the welfare issues of the people. Once there is a certain degree of comfort [and] they have been exposed to schools, hospitals, good roads, health services ... I think we could see some reduction in corruption.

Respondents were concerned about the structural challenges facing citizens and officials; however, this is not to say that they did not also reflect the mainstream view. A senior employee acknowledged the positive nature of PNG's culture of 'give and take', but suggested that, ultimately, leaders need to obey the laws and bureaucratic procedures of the land:

People seek refuge in corruption to explain that this kind of sharing is done in the culture, and that [way of thinking] sometimes cuts across, you know, cuts across laws ... The Ombudsman Commission operates under two Organic Laws and often when you have referrals of MPs one of the things they always [say] is 'look we are basically responding to what people would normally ask. You know the electorate, our voters, are asking for this and what are we supposed to do? You are handicapping us by waving all these legal impediments, while at the same time we, as people who are elected in the role of *big man* and chiefs, have to deliver.' I mean that's fine if you are sharing something, which everybody contributes to ... But I think when you translate that into the nation-state obviously you'll have to have rules, regulations and laws.

As this quote shows, common law was a crucial touchstone for addressing corruption, but its importance is discussed in relation to its relevance to local cultures. Thus, while the mainstream view's legal and public office perspectives were ultimately seen as important for addressing corruption, concerns about how cultural values fit into this mix were still important.

In summary, the OC PNG has struggled to address corruption in the ways that the CPC and Narokobi envisaged, yet the moral concerns of PNG's founding leaders still resonated through the concrete and steel of the notorious Deloitte tower into the OC PNG's meeting rooms and offices. Respondents articulated a longing for a more meaningful engagement with the alternative view on corruption beyond what the OC PNG was mandated to provide. In some ways this is a striking response given that the OC PNG is, officially at least, so focused on providing a legal response to corruption. Many interviewed drew on their day-to-day interactions – which included investigating allegations of corruption by politicians and other state officials – to suggest that structural issues were often at the heart of PNG's corruption problems. Respondents believed that culture, poverty, poor pay, poor education and lack of understanding significantly contributed to corruption. Away from the OC PNG's official documents about corruption was a sympathy for those in public office that turned corruption into a very grey concept; in the words of those associated with the OC PNG, corruption was a very different problem than suggested by the country's laws and the OC PNG's legal mandate.

The Coalition

PNG has a history of street-level protest. In 1990, for instance, students, union members and others joined in the streets of Port Moresby to protest against growing levels of corruption (*PNG Post-Courier*, 1990a, 1990b). In 1997 civil unrest towards the government's decision to hire mercenaries to fight a civil war in the island province of Bougainville brought the capital Port Moresby to a standstill, and helped bring about the downfall of a Prime Minister. More recently, in 2016, university students across the nation boycotted classes in response to PM O'Neill's decision to sideline the head of the nation's fraud squad, and due to accusations of corruption involving the PM himself. Walking the streets of the capital, Port Moresby, often reveals street preachers preforming to small crowds. Some of these preachers are religious, speaking about the need to repent and impressively reciting long tracts of the Bible from heart. Others have engaged in political protest, which has occasionally spilled over into nationwide conflagration. This section focuses on a group of local anti-corruption protesters called the Coalition, who were vocal in their street demonstrations. It examines how the Coalition were different from their predecessors, a radical group known as Melanesian Solidarity (MelSol). In doing so, this section highlights how one local group of activists have interpreted both the mainstream and alternative views on corruption. The presentation of the material in this section is different to how findings are set out in other chapters of the book where anti-corruption organisations are examined (in this chapter and the next). Here, because the differences between official and unofficial narratives are not as stark as for other organisations, the Coalition's 'on-stage' and 'off-stage' narratives are presented together.

Background

The Non-Government Organisations and Civil Society Coalition (referred to as 'the Coalition', a shorthand term used for the purpose of this book) was a loose alliance of four registered local NGOs: Eda Hanua Moresby Inc. (EHMI), PNG National Awareness Front (PNG NAF), PNG Millennium Good Governance Organisation (PNG MGGO) and National Capital District Youths Association (NCD YA) (PNG National Awareness Front et al., 2007). The group formed in 2007, and since that time conducted public protests – at the time of fieldwork they were directed towards the Somare government; however, in recent times the group has also criticised his replacement, Peter O'Neill – in public places in Port Moresby. In 2008–09, the Coalition mostly raised funds for its activities during protests, with leaders receiving between 160 and 170 kina each. Other sources of funding were sought from local businesses and anti-corruption actors in PNG, but the Coalition reported little success in these endeavours. Having said this, no written records of accounts were kept, and there were no official lists of members, although the group was keen to claim that tens of thousands were associated with the group. As with many grassroots movements in PNG and elsewhere, they found it difficult to keep accounts. This constraint came with a dose of irony, given how passionately the Coalition called for measures to improve transparency and accountability within government.

I first met leaders of the Coalition out the front of a local fast-food outlet, Big Rooster, in the suburb of Boroko, a suburban hub known as 'four mile' because it is four miles from Port Moresby town centre. We first met in November 2007 and chatted over take-away chicken and chips. I was told that some of the group's leaders had just come out of hiding after unsuccessfully attempting to march on Parliament House to present a petition calling for Somare's resignation. Their attempt was met with threats from the Chief of Police, who suggested that the group had threatened the PM's life; fearing reprisals, they took a break from campaigning. Despite this push back, by the time I first met them, they were keen to return to the streets and highlight what they saw as increasing corruption from the Somare government. And a few months later they were back, again in the suburb of Boroko but this time with a hand-held megaphone in tow, bellowing out all the reasons why PM Somare had to go.

Over time I got to know these activists, and I moved with them from their sites of protest to their headquarters and homes. They worked out of a more austere location than their anti-corruption counterparts. Their 'headquarters' was a fibro house on stilts, located in the outer suburb of Gerehu, a suburb that has become notorious for its violence and robberies. It's a place where expatriates and some Papua New Guineans fear to tread. While they would have liked to have had more money at their disposal, the Coalition were proud of their austere surrounds and resources. Comparing themselves to other NGOs working in the country, one leader was proud that the group were able to live among and like PNG's *grasruts*. A few voiced disdain for 'expensive cars', the salaries and the meetings in 'air-conditioned hotels' which were associated with other NGOs that were supported by international donors.

Interpreting and protesting corruption

The Coalition were au fait with the importance of referring to an obligatory definition of corruption, in line with policy documents of government agencies and international anti-corruption organisations. In their most comprehensive 'official' document, 'The 31st Independence Anniversary Petition Against Corruption', The Coalition defined corruption as: 'people with authority or power willing to act dishonestly or illegally in return for money or personal gain', a definition attributed to the 'Oxford Advanced Learner's Dictionary' (Millennium Good Governance, 2006: 20) – essentially the 'entrusted power' definition explained in chapter 2. In public events the Coalition also prefaced their oratories with their definition of the concept, sometimes repeating this definition to clarify for the crowd what they meant when they went on to accuse the government of gross corruption.

Moving from this expansive definition, the Coalition highlighted numerous causes of corruption: from poverty and the cost of high living through to 'power hungry' and egocentric leaders and bureaucrats (Millennium Good Governance, 2006: 20). But there were three key causal factors that went to the heart of the Coalition's concerns about why PNG was suffering from corruption. The most important cause of corruption for the Coalition, as previously noted, was PM Somare. For the Coalition the then PM was not just making poor decisions leading to corruption; he himself was an embodiment of corruption. At a protest rally on 5 June, 2008, after protesting against the corruption of the Somare government, one of the Coalition's leaders told the crowd: 'corruption started in 1975', the date Somare first became PM when he led the country to independence. In describing the PM's wrongdoing leaders often portrayed Somare's actions in terms of a personal violation. When asked about what the key purpose of the group was, one of their leaders said they were trying to 'assist the Papua New Guineans to make sure that [the PM] does not stay in power and that he does not rape the nation and our children's future'.

The Coalition cited numerous concerns about the then PM's corruption. These were contained in an evolving petition that grew in length while I was hanging out with the group; the list of illicit activities and illegalities grew to over 60. One iteration suggested that in order 'to save the nation from further ruin' Somare must step down, because he had:

- Weakened the powers of the OC PNG to investigate corrupt leaders;
- Fought against the OC PNG in court to halt investigations into allegations of misconduct in office;
- Failed to come out publicly and speak against allegations of corruption involving civil servants and ministers in his government;
- Failed to discipline MPs allegedly involved in corruption;
- Engaged Pacific Register of Ships, a company in which Somare allegedly owned shares, to perform statutory vessel certification, ship safety and survey services on behalf of the National Maritime Safety Authority (this represented, to the Coalition, a conflict of interest);
- Appointed his son, Sana, and daughter, Bertha, to senior positions in government (adapted from Kolao, 2008).

As a result, the Coalition maintained that 'under Somare's leadership, corruption has increased at [the] fastest rate' (Millennium Good Governance, 2006).

These views were channelled into rage-filled sermons which rallied supporters to join them in calling for the PM's resignation. In place of the PM they called for the installation of one of four government Ministers (all of whom were a part of Somare's government in 2008), namely:

- Dr Puka Temu (Central Province), Deputy PM and Minister of Lands, Physical Planning and Mining;
- Don Poyle (Enga Province), Minister for Transport and Civil Aviation;
- Patrick Pruaitch (West Sepik Province), Minister for Treasury and Finance;
- Paul Tiensten (ENB Province), Minister for National Planning & District Development.

Support for these politicians was no indication of their lack of corruption. Pruaitch was referred to the public prosecutor for corruption in 2009. In 2014

Tiensten was sentenced to nine years' imprisonment for corruption; in 2015 an additional three years were added to his sentence for another corruption-related crime. During the time of fieldwork, the Coalition's support of these politicians produced discord among onlookers; in interviews some suggested that the Coalition were too politically partisan, and were likely paid by the politicians they supported.

The second target of the Coalition's scorn was the growing community of Asian traders (mostly Chinese), whom they saw as both directly and indirectly causing corruption. The Chinese community in PNG has a long history, which can be traced back to the nineteenth-century German colonialists who brought in Chinese as labourers to work on plantations. Today, Chinese in PNG can be heuristically divided into two camps: 'old Chinese' who lived through colonisation, and the more recently arrived immigrants sometimes referred to as the 'new Chinese' (Chin, 2008). While the two groups are sometimes conflated in popular discourse, it is this latter group who are often labelled 'illegal' immigrants and as threatening Papua New Guinean livelihoods. The new Chinese – perhaps more so than other Asian communities such as the Malay community – have become a target for popular discontent about their role in the country's economy and society.

Some in PNG fear that Asian traders are able to set up trade stores and other businesses across the country, and have taken away opportunities for locals to engage in commerce. There are an increasing number of Chinese-operated trade stores in remote villages. One reason for the spread of these stores is because Papua New Guinean traders find it difficult to run a profitable business when their *wantoks* (relatives, friends and clan members) expect to be given stock for free. The second concern expressed in PNG (as Smith, 2012 highlights) is the nature of large-scale Chinese investment in the country. The Ramu Nickel Mine in the northern province of Madang is an example, with the Chinese state-owned company accused, particularly by those directly affected by the mine, of locking locals out of employment opportunities (Smith, 2012). A large multi-industry Malaysian company run by ethnic Chinese businessmen, Ribinaun Hijau, has also been associated with corrupt and unethical practices in PNG. The company has been implicated in corruption scandals in the forestry sector over the past three decades (Barnett, 1989; Greenpeace, 2005; Smith, 2012).

Reflecting populist discontent, the Coalition was often eager to let people know about their rage towards this community. In a press release they rallied against Asian migrants stealing Papua New Guinean jobs, selling 'counterfeit and unhealthy' products and entering the country illegally to work for Chinese firms (notably the Ramu Nickel Mine). These perceived slights, in the minds of these activists, helped perpetuate corruption. In their policy for good governance, they argued that these issues resulted in the 'high cost of living [which] encouraged people to find other means of generating money such as corrupting [the] public purse' (Millennium Good Governance, 2006: 20). In this logic, if citizens did not have access to jobs, 'proper' goods and economic opportunity, corruption would prevail.

The Coalition was particularly concerned about the spread of Asian businesses throughout the country. While a few admitted that it was difficult for Papua New Guineans to operate trade stores given the expectations placed on them by *wantoks*, others were certain that these groups were successful because of their ability to, in the words of one leader, 'bribe government officials'. For the Coalition, Asian businesspeople were engaging in a range of corrupt practices around the country and could get away with it because of their relative economic power.

As a corollary, the Coalition rallied against Asian shop-owners, which resulted in an anti-Asian rally that spread across the country. Initially held in Port Moresby on 14 May, 2009, the protest spread to Lae, Goroka, Madang and other regional centres, where thousands of locals rioted, looted local shops and destroyed property. Three looters were shot dead, one was trampled to death (Callick, 2009), while three ethnic Chinese died (Smith, 2012). The damage was extensive and resulted in a parliamentary enquiry. And in the middle of this carnage was Noel Anjo, one of the Coalition's key leaders. In Port Moresby he led the protesters, and had his views heard through local and international media. On New Zealand radio he explained what he saw as the rioters' demands:

> We want investors but we don't want robbers. We see the Asians as robbers, not investors. So they don't have any respect for our laws, they don't have any respect for our customs, for our people. They are mafia. We don't need them. So we have given the government as of 2009 to clean up the country and remove unwanted Asians from this country. And as of 2010, the first parliament session, we want a response from our government regarding our position.
>
> (Radio New Zealand International, 2009)

In the same interview Anjo expressed concern about the relative economic power of Asian businessmen and employees, with Radio New Zealand noting that he said 'the public is fed up with the Asian dominance of PNG's businesses and jobs which, he says, should be reserved for local people' (Radio New Zealand International, 2009).

The anger that some in the Coalition felt towards the Asian community was palpable even before the riots broke out. In the city's inner suburb of Boroko in late 2008, one of their leaders spoke of his burning passion: to take back power from the Asians whom he believed had taken over the country. As we strolled through a dimly lit shopping mall he pointed out the small shops selling cheap electronic appliances, footwear and other imports, owned or run by Asian workers. 'I want all of these people out, they sell bad products,' he said. For the Coalition then, curtailing the economic power of the Asian community – especially Asian traders – required their expulsion from the country. This would, in their logic, limit the ability of Asians to directly or indirectly promote corruption and other social ills.

For the Coalition, the malignant influence of Western modernity was the third key cause of corruption in the country. A volunteer with the organisation noted that corruption was occurring because:

> People are trying to live a life with the Western style of living. You have the shop prices; the food prices in the shops are going up, so that's why the other people try to match up with the others. That's why they have to make more money [and] try to live in bigger cities like Port Moresby, Lae, Mt. Hagen, Goroka and Madang.

Similarly, in one of their policy documents the group somewhat poetically outlined their concerns about the pernicious impacts of Western culture on Papua New Guinean political life:

> The echo of modern songs and music beats in the nightclubs [...] easily [penetrates] into the walls of our Parliament House and beats persistently into the hearts of our members making them dream and wonder for more. Thus the foreign voice of corruption deafens their ears resulting in [them] ignoring 'PNG Ways' and other National Goals and Directive Principles that are set in the Preamble of our National Constitution some 30 years ago.
>
> (Millennium Good Governance, 2006: 10)

Some in the Coalition believed, as one leader pointed out, 'white men' have 'taught Papua New Guineans how to be corrupt'.

Despite the group's at times nationalist rhetoric, the Coalition did not believe that Papua New Guineans had the wherewithal to deal with corruption outside of showing their dissent through protests. Instead, the Coalition lauded outsiders' approaches to fighting corruption. For example, during a rally held in the Port Moresby suburb of Gordons, in July 2008, one leader argued that Western approaches to fighting corruption should be mimicked, saying:

> Papua New Guinea, we are pretenders ... where the white people come from, when they talk about ending corruption and removing corruption, they action it. Papua New Guinea, in order to end corruption, we do not know how to action it.

To enforce penalties against those found to be corrupt, the Coalition advocated for 'independent foreign law enforcement agencies like [the] Australia ECP [Enhanced Cooperation Programme] to be in charge of the corruption Court and Tribunal cases' (Millennium Good Governance, 2006: 7).

This programme was certainly not a nationalistic response to corruption. Announced in December 2003, the ECP was 'designed to provide long-term, hands-on, technical assistance from senior and highly skilled Australian government officials to key economic, justice and border security agencies in GoPNG' (Dixon, Gene, & Walter, 2008: 9). After the Supreme Court of PNG ruled that the conditions under which Australian police were deployed were unconstitutional in 2005, a scaled-back programme commenced in 2006 with police advisors rather than inline police offices. In endorsing the ECP, the Coalition were supporting foreign intervention of a particular type. While they rallied against Asian influence, they welcomed, with open arms, Western-backed institutions. While they expressed concern about the corrupting influence of Western modernity on the country, the Coalition did not advocate for Westerners to leave. In fact they called out to 'white people' to save the country from the disease that they had introduced. In doing so, they embraced the market-oriented logic of the mainstream view's economic perspective.

Ghosts of the past

The Coalition's response to corruption represents a distinct break from their immediate predecessors, an NGO called Melanesian Solidarity (MelSol). Launched in 1984, MelSol were also ex-University of PNG anti-corruption activists who were based in the settlements of Port Moresby. Like the Coalition they spread their messages through street protests, and on occasion clashed with police. And the two groups were connected by a local politician – Powes Parkop – who was General Secretary of MelSol and helped formed one of the NGOs associated with the Coalition. Yet the two civil society groups differed in one key aspect: their political orientation.

The Coalition called for their ex-colonial masters and the private sector to come in and fix corruption, but years before MelSol was rallying against what they considered neo-liberal and neo-colonial intervention (Walton, 2016). They directed their concerns towards Australia's aid programme. In 1998, Parkop condemned Australia for supporting the structural adjustment policies of the World Bank and IMF (Dixon, 1998). Later, in 2003, Parkop suggested that the Australian aid programme in PNG was ineffective, and that Australia was partly to blame for policy failures of the PNG government (Radio New Zealand International, 2003). MelSol also had regional powers and international financial institutions in their sights, with Parkop proclaiming at a conference in Sydney, Australia, in 1998 that colonialism in the Pacific took three forms:

> 'old European colonial rule' [that] continues in countries such as Tahiti (French Polynesia) and Kanaky (New Caledonia); 'south-south colonialism' by Indonesia in East Timor and West Papua; and the 'recolonisation' of the region by the World Bank, the International Monetary Fund and transnational companies.
>
> (Dixon, 1998)

The politically left-leaning MelSol projected their voice through like-minded overseas newspapers: they were featured in the UK newspaper and website

Socialist Worker (published by the Socialist Workers Party) and the somewhat similar Australian socialist newspaper, *Green Left Weekly* (published by Socialist Alliance). As a result, MelSol sought to fight corruption without strengthening the hand of multinationals and other foreign institutions.

The 1990s and early 2000s were a time of great upheaval in PNG. It was a time of public protest involving a range of civil society organisations – including NGOs and churches – and often led by MelSol. In 1995, citizens protested against an IMF/World Bank proposal to convert communal land into alienable assets (in the hope that such reform would make it easier for businesses to invest). Two years later, the PNG government's failed attempts to hire foreign mercenaries to fight its then long-running war in Bougainville – known as the Sandline Crisis – brought citizens and the military out on the streets in protest. The scandal highlighted secretive deals between the government and the mercenaries. For MelSol the protests were about fighting corruption; Michael Tataki, an executive member of MelSol, said at the time that 'The overriding issue for ordinary people is corruption' (Dixon, 1997b). He went on to explain that MelSol played a key role in educating students about the corruption involved (Dixon, 1997b).

MelSol was again out on the streets during the height of the privatisation programme in 2001, when thousands of students and citizens rallied at Parliament House in protest against the threat of privatisation of government services including Telikom, Elcom and Post PNG. The protest tragically resulted in the deaths of three students from police bullets (Beder, 2001). But the independence of the movement was severely challenged when some of its leaders went into politics. Following Sandline, 13 MelSol-affiliated candidates were elected on an anti-Sandline and anti-corruption platform (Griffin, 1997: 77). Bill Skate became the new PM with support from pro-MelSol independents. The Skate government was plagued by political scandal, which resulted in the Catholic Archbishop of Port Moresby, Brian Barnes, calling for a change of government and the establishment of an Independent Commission on Corruption (Standish, 1999). The ensuing controversy placed MelSol's activists-turned-politicians in a poor light, and resulted in criticism from their supporters (Dixon, 1997a).

In 2007, MelSol was further weakened when Powes Parkop, then MelSol's General Secretary, campaigned for and won the governorship of NCD, which covers the capital Port Moresby. By becoming a part of the National Parliament, rather than opposing it, Parkop put his efforts into governing his electorate, and MelSol's voice was diminished. The organisation has not played a significant role in public protest since the early 2000s.

At the time of writing, NCD Governor Parkop has moved away from progressive politics. Some of the concerns that burned bright during his MelSol days are still apparent: while it took him some time after his election to speak out, Parkop has voiced his support for West Papuan leaders, who have faced oppression and human rights abuses from the Indonesian armed forces since 1967, the year that Indonesia forcefully colonised the region.

Many in the previous Dutch colony have a strongly held desire for independence. Still, Parkop has openly supported policies that sought foreign investment through privatisation. In 2014, while Parkop was visiting Canberra, he spoke about his support for privatisation schemes while governor. 'Stopping privatisation and going to corporatisation has probably not worked out ... we can't keep supporting PNG Pawa and Air Nui Guinea [state-owned enterprises] forever,' he said. I asked Parkop why he thought there are fewer progressive civil society organisations now than in MelSol's heyday. He replied that PNG is now in a very different historical moment: 'PNG [needs to and] is trying to sort itself out for the future and find what entry points it has in the global community,' he said. In other words, he believed it was time for PNG's civil society groups to move on from their left-leaning politics that challenged privatisation and other neoliberal policy prescriptions, and support PNG's full integration into the global capitalist economy.

Thus, the Coalition's reactionary activism reflects a broader move away from left-wing progressive politics in the country. It suggests that some parts of civil society in PNG are moving further away from the CPC and Narokobi's moral and critical concerns about the nature of development. While the CPC stressed the importance of maintaining Papua New Guinean Ways in dealing with modern problems – unlike MelSol – the Coalition did not believe that Papua New Guineans had the wherewithal to fight corruption. Similarly, while the CPC's final report and MelSol rallied against corruption in developed countries, the Coalition saw corruption as a distinctly Papua New Guinean and Asian problem. For them, developed countries, Australia in particular, were PNG's new anti-corruption saviours. The story of the Coalition then is a story of broader civil society's jump to the right, and its loss of faith in the promises of Papua New Guinean Ways.

Discussion and conclusion

The tension between modernity and tradition is a central part of PNG's history, society and politics. This tension permeates activities and discourse of national anti-corruption institutions. It is almost impossible to mention the word *korapsen* (corruption) among those who engage in national responses to corruption in PNG without encountering difficult conversations about the trade-offs between culture and modernity. Those from the OC PNG were concerned that the answers offered by the mainstream view of corruption – i.e. responses tied to the state – were limited. They not only sought a more nuanced understanding of the causes of corruption than state-oriented approaches allow, they also longed to see a more robust Papua New Guinean response to the problem. What this would look like is still fuzzy, but the desire remains.

However, over time the possibility for developing culturally relevant responses to corruption has been degraded and disowned. Attempts to formalise the alternative view within the OC PNG have failed, with politicians and the authors of the constitution effectively defanging this institution and giving it a more mainstream state-oriented outlook than the CPC envisaged. The failures of the OC PNG have been accompanied by an increasing hostility towards the Papua New Guinean Ways and the Melanesian Way, with a number of commentators considering these concepts a convenient excuse for elite corruption. The Coalition's degrading of Papua New Guineans' abilities to fight corruption and their calls for foreign intervention present the apogee of this national degradation of alternative views on corruption and, by extension, development.

In rejecting the left-leaning politics of MelSol – a politics that reflected the core tenets of the Papua New Guinean Ways – the Coalition reflect a broader discontent with the ability of the PNG state to provide development. The dilapidated government buildings in the capital and around the country lie in stark contrast to the gleaming private spaces that have sprung up around the capital, Port Moresby. The state's standing has been further tarnished by its poor development track record and reputation for corruption. This has meant that, in the name of anti-corruption, the dreams of Narokobi and the CPC have been thoroughly rejected by this group. For them, the tidal wave of modernity that was so feared by the CPC is now needed to wash away PNG's failures.

Thus, supporters of the alternative view on corruption were found where they are least expected. It is those within the legally mandated OC PNG who most strongly reflected the alternative view, while the Coalition most strongly rejected this view. In part this reflects the differences between a middle-class nationalism and the more negative nationalism that can frame local Papua New Guinean responses to Western institutions. This middle-class nationalism is in part fostered because the middle-class are less likely to feel the direct impacts of state failures than those who are more marginalised by development (such as the Coalition). In turn, the potential for more culturally aware anti-corruption responses shaped by Papua New Guinean values reflects the dreams and aspirations of those who are heavily invested in the state working – as those connected to the OC PNG are.

The Coalition's lost faith in the state and fellow citizens to deliver development, reflects what anthropologist Joel Robbins calls 'negative nationalism'. Robbins' study of the Urapmin people of West Sepik Province (in the northwest of PNG) found that, influenced by Christian teaching, the Urapmin considered themselves sinners, inferior to members of other nations, such as Australia or the United States. The Urapmin thought of their Papua New Guinean nationality as a negative attribute:

> the Urapmin see their national identity as Papua New Guinean as a source of much that they dislike and wish to reject in themselves.
>
> (Robbins, 1998: 110)

Robbins called this 'negative nationalism' – the product of being constantly reminded of the inferiority of one's nation. The Coalition's concern about corruption, and the failure of the PNG state both contributed to their sense of negative nationalism, and meant that they did not believe that Papua New Guineans could respond to their own problems. As I've argued elsewhere (Walton, 2016), the Coalition's response (and the reason it is different to MelSol's) is also caused by their lack of engagement with transnational development actors. While MelSol was connected to transnational NGOs and political movements the Coalition's outsider status had then focused inwards, towards Asian traders and their leader's corruption. Comparing the Coalition's engagement and understanding of corruption to the OC PNG shows that the hope of PNG's founding leaders lives on in enfranchised pockets of society who are still waiting for the promises of a strong, autonomous and culturally aware state to materialise.

5 Integrity warriors
International anti-corruption agencies' views on corruption

Introduction

Over the past two decades anti-corruption measures have become a standard part of most policy documents, and funding for anti-corruption activism has increased exponentially. The growth of anti-corruptionism has resulted in an interconnected and globalised anti-corruption industry, whose members share ideas about corruption, and convert some of these ideas into programmes, funding opportunities and policies. This discourse, as shown in chapter 2, is reflected by the widespread support – at the international level at least – for the different permutations of the mainstream view on corruption.

But how influential is this industry's discourse on corruption? How well does it move from the international realm of international policy documents, websites and media statements to the messy realities of national anti-corruption responses? What perspectives of corruption are reflected by representatives of this industry, both in the public sphere and in 'off-stage' communications? Is the mainstream view always favoured over the alternative view? In this chapter we will explore these questions in relation to PNG's two most prominent international anti-corruption organisations: Transparency International PNG (TI PNG) and the Australian aid programme (once known as AusAID).

The chapter argues that international concepts about corruption have been woven into the fabric of both organisations. Yet there are enough holes and imperfections in this fabric to challenge the notion that these organisations are cut from the same cloth. In particular, it is argued that both TI PNG and AusAID only partially reflect international perspectives on corruption. I suggest that within the international anti-corruption industry, perspectives on corruption are transferred unevenly, although ultimately the mainstream view dominates framings about corruption and anti-corruption responses.

The chapter first examines the rise of the international anti-corruption industry and provides a background on the arrival of international anti-corruption organisations in PNG. It then introduces TI PNG. After providing a background on this international NGO it examines the organisation's official and unofficial narratives about corruption. The Australian aid programme (then known as AusAID) is next. This section examines the 'on-stage' (official) and 'off-stage' (unofficial) communications around corruption. The conclusion discusses what these findings mean for understanding the nature of anti-corruption efforts in PNG, particularly in relation to the framework developed in chapter 2.

The global anti-corruption industry

The mainstream view on corruption is now the orthodoxy of the global anti-corruption industry (see chapter 2). However, this has not always been the case. Indeed, international discourse on corruption has changed, quite dramatically, over the past few decades. Understanding this change requires travelling back to the 1960s, a time when many Western governments and academics had an ambivalent attitude to corruption, accepting its existence as an ineluctable consequence of modernisation. This ambivalence was, in part, connected to the political manoeuvrings of the Cold War; for both blocs, securing the political support of a country through aid was far more important than monitoring the utilisation of it. As a result, the West, along with the rest of the world, was not outwardly concerned with corruption. Indeed, up until the 1990s the word 'corruption' was noticeably absent from the lexicon of large development institutions, such as the IMF and the World Bank (Krastev, 2004). It was not only that the 'c-word' did not officially exist; there was also some serious consideration given to the potential benefits that corruption could bring to developing countries (Huntington, 1968). Development actors had thus not yet embraced the mainstream view on corruption, indeed their policies and programmes suggested that corruption was not a central problem for development.

With the end of the Cold War, corruption came to be perceived as a key impediment to development, rather than an outcome of it, and in turn the mainstream view became a part of international development discourse. It was during this time that advocates of market liberalisation found anti-corruption policies to be a useful means of overcoming protectionism (Krastev, 2004). The reasons for this change are contested. Some contend that Western donor agencies only started to worry about corruption when it became evident that their own policies were exacerbating poor social, economic and environmental outcomes. Szeftel (1998) has argued that international donors blamed poor socio-economic outcomes in Africa on government 'corruption' instead of accepting responsibility for their own failed structural adjustment policies. But the story that has most purchase, within and outside of the industry, is that of a small group of development officials, who came to realise that the pernicious nature of corruption had to be addressed if conditions of the poor and marginalised were to improve (Hope & Chikulo, 2000; Reilly, 2000; Mbaku, 2007). It was the experiences of mostly ex-World Bank employees, who were frustrated with the levels of corruption in Africa, which led to the formation of the most influential anti-corruption organisation in the world – Transparency International (TI). Formed in 1993, the organisation's Secretariat is based in Berlin and at the time of writing had more than 100 locally established national chapters.

As a global 'movement', TI drives much of the global anti-corruption agenda (Hope & Chikulo, 2000; Krastev, 2004) and monopolises messages about corruption in the countries in which it is based (Larmour, 2005).

Pressure from groups such as TI has led to major Western donors prioritising anti-corruption programmes and policies. In 1996, after much campaigning by TI, the World Bank committed to combatting corruption through measures ranging from public sector reform to debt cancellation (Cartier-Bresson, 2000: 11). Eliminating corruption has been a key concern of the World Bank ever since (World Bank, 1997, 2007). With the World Bank on board it was only a matter of time before other major Western donors joined the cause. Now, most donor organisations have an anti-corruption policy of some description. The international anti-corruption industry, with support from Western governments, has become a tour de force – influencing international, regional, national and sub-national organisations, institutions and citizens.

The international anti-corruption industry is now composed of a conglomeration of NGOs, aid donors, financial institutions, think tanks and private sector organisations, all attempting to mitigate corruption. The industry has grown exponentially over the past decade, with one study estimating that spending on anti-corruption activity increased by 50 times during 2003–2009 to be worth US$5 billion (Michael & Bowser, 2009: 1). This has meant anti-corruption has shifted from a movement to a very powerful industry, over a short period of time (Sampson, 2010). In terms of research there has also been substantial growth. The causes, consequences, effects and nature of corruption have become a favourite topic of political scientists and economists. Think tanks and universities offering courses on corruption have increased. Michael and Bowser (2009: 1) estimate that 9 English-language academic authors wrote extensively on *anti-corruption* in 2003; by 2009 almost 70 academics were conducting research on the topic. Google Books' Ngram Viewer shows that mentions of the word 'corruption' in books in its depository increased by 48 per cent between 1990 and 2008.

Krastev (2004) puts the successful spread of the anti-corruption industry down to the transnational character of TI, which gained legitimacy through setting up national chapters around the world and in turn could speak with a local voice. In turn, Bukovansky (2006) argues that TI is more likely to meaningfully address corruption because it better represents the concerns of developing countries than other development actors. She contrasted TI's approach to that of the World Bank and the IMF, whom she considered as bringing a top-down approach to governance by providing loans to developing countries with strict conditions around governance reforms attached. For these scholars, the industry becomes more legitimate the more it connects to local societies (see Walton, 2016).

The anti-corruption industry comes to PNG

As noted in the previous chapter, there is a long history of concern about corruption and responses to it from state-based institutions and activists in

PNG. International organisations have become a key part of both promoting the problems of corruption and responding to it. The rise of international anti-corruption organisations can be traced to the late 1990s, when local events and international actors helped amplify narratives about corruption across the country. Concern about corruption increased after the government's failed attempt to hire foreign mercenaries to fight in its long-running war in Bougainville, a mineral-rich island province off the east coast of the mainland. This event was beset by accusations of backroom deals and became known as the 'Sandline Affair' (after Sandline International, the name of the company that provided the mercenaries). At the height of the crisis the then Governor General, Sir Wiwa Korowi, placed a full-page advertisement in both national newspapers warning that the crisis showed 'Papua New Guinea will be destroyed by ... greed, corruption, nepotism, malice, selfishness and manipulation' (Dorney, 2001: 154). The Sandline Affair ultimately caused the downfall of the then PM, Sir Julius Chan, in 1997. The scandal convinced many in the public that 'their leaders were corrupt' (Dorney, 2001: 154).

The same year, 1997, saw international organisations start to significantly contribute to the fight against corruption. This was the year that TI commenced operations in PNG. Since its inception, the agency has been at the forefront of anti-corruption campaigning (Larmour, 2003), along with other civil society organisations and the media. As a result of TI PNG and its partners' efforts, corruption has become one of the most discussed topics in the country. Also in 1997 Australia, the largest international donor to PNG, began taking the good governance – often a euphemism for anti-corruption (see chapter 2) – agenda seriously, with the Australian Department of Foreign Affairs and Trade (1997: 15) acknowledging that 'governance was a particular focus of policy development during the year'. Subsequently, funding to PNG for governance programmes has increased considerably, as has discussion about corruption.

While national organisations and events have played an important role in responding to corruption (see chapter 4), international organisations – particularly TI PNG and the Australian aid programme – through their funding and advocacy have also contributed to the way in which Papua New Guineans understand and respond to corruption. The following section examines the most vocal of these international organisations: TI PNG.

TI PNG

At the time of my fieldwork (2008–09), TI PNG was located in the Mogoru Moto building in downtown Port Moresby. The building rose up from the beetle nut-stained streets to look over Fairfax harbour. TI PNG's offices were situated among those of lawyers, investment advisors and insurance agencies. Comprising mostly open-plan office space, the organisation's office was tightly fit together. With office space at a premium in the years of the LNG boom, staff and numerous volunteers crammed into interconnected desks shaped by office partitions and bookshelves.

In those offices staff worked through the day and sometimes through the night to deliver on the agency's expanding number of activities. In 2008 the organisation's work revolved around five overarching strategic priorities: advocacy, civic and electoral education, memberships and building coalitions, legislative initiatives and policy research (Transparency International Papua New Guinea, 2009b). These strategies directed the agency into a range of projects, which included:

- Conducting the second annual Walk Against Corruption event in both Port Moresby and Kokopo (the capital of East New Britain province);
- Holding a youth democracy camp and subsequent radio shows;
- Hosting a forum for Pacific women speakers;
- Conducting research into Papua New Guinean perceptions of corruption;
- Piloting the development of resources for an anti-corruption civic education project.

Such activities have helped to further build the agency's profile and raise awareness about corruption in PNG.

As a 'chapter' of its parent organisation, TI, in January 1997 TI PNG started operations involving Papua New Guineans led by the late lawyer, politician and public servant Sir Anthony Siaguru. The organisation's initial goal was to combat corruption, and promote openness, honesty and accountability in public and private dealings (Transparency International Papua New Guinea, 2009a). The organisation's vision for PNG was 'an independent nation in which government, politics, business and civil society and the daily lives of our people are free of corruption' (Transparency International Papua New Guinea, 2009b: 6). Its stated mission was to 'reduce corruption and create a better future for our children by:

- Promoting honest and good leadership;
- Supporting active public involvement in the struggle against corruption, and
- Informing and educating the public on anti-corruption culture'. (Transparency International Papua New Guinea, 2009b: 6)

The organisation stressed that these goals were not just the aspirations of the organisation, but tied to the will of PNG's citizens. The organisation's 2007 Annual Report stressed that the agency's 'performance is evaluated by the continued relevance and content of key messages and programmes delivered to [TI PNG's] beneficiaries – the people of Papua New Guinea' (Transparency International Papua New Guinea, 2008a: 6). In doing so, TI PNG identified PNG citizens as key constituents who legitimise their operations; an approach that Krastev (2004) suggests helped legitimate and spread the messages of the anti-corruption industry around the globe. As we shall see, this is in marked contrast to the Australian aid programme, who have sought to legitimise their work in PNG by linking it to the priorities of the PNG government. With Papua New Guineans on the board of management and employed by the organisation, TI PNG was setting itself up as a transnational organisation whose activities were guided by international anti-corruption discourse and grounded by concerns of PNG's citizens.

By the time I had arrived in 2008, TI PNG had grown significantly with support from a mix of development actors, businesses and citizens. Between 2005 and 2008 TI PNG's income almost doubled, increasing from 535,327 kina (US$168,360) to 1,061,420 kina (US$333,817). By 2008, most funds were contributed by donors for specific projects, as opposed to core untied funding. Major contributors included New Zealand Aid, the British High Commission, TI Secretariat, the US State Department, the European Commission in PNG, Save the Children in PNG, Oxfam International, Air Niugini and Pacific MMI Insurance. Membership of the organisation required a nominal fee of a minimum of 5 kina for individuals. In 2008, 131 individual members gave a total of 3,425 kina (US$1,077), while 33 corporate or organisational members contributed 30,899 kina (US$9,718) (Transparency International Papua New Guinea, 2009b: 22). In total, only 0.3 per cent of the organisation's funds came from individual members – the rest came from businesses, as well as international aid organisations and NGOs.

Official definitions, causes and responses

The TI Secretariat has by and large eschewed relativist definitions of corruption. Early on in the organisation's history it rallied against cultural interpretations of the concept (Pope, 2000b). More recently it has been less antagonistic, although it still defends a definition of corruption that is supposed to provide an interpretation of the concept across cultures. TI's website has rather defensively stated that its definition – 'the abuse of entrusted power for private gain' – provides 'a clear and focused definition of the term'. While TI points to this as a universal definition, as discussed in chapter 2, it has also broadened the way the organisation defines the term since 2000 when it adopted the 'abuse of entrusted power' definition. It asserts that the legitimacy of its most recent definition is confirmed by its 'Global Corruption Barometer survey, which analyses people's views and experiences of corruption in more than 60 countries' (Transparency International, n.d.-a).

TI (TI PNG's parent organisation) has sought to address any critique of this approach by suggesting that its national chapters will respond to concerns about cultural interpretation. In a reluctant acknowledgement of the numerous cultural interpretations of corruption, its website has noted that:

> The forms and causes of corruption vary across countries, however, meaning that the best ways to address it differ too. This is why our

approach to fighting corruption is grounded in our system of national chapters, which are run by people who are anchored in their societies and are therefore in the best position to understand and tackle corruption in their respective countries.

(Transparency International, n.d.-a)

The chapter system – which gives country chapters a degree of freedom in interpreting and fighting corruption – allows TI to both present a universalistic stance and demonstrate its local, culturally grounded, credentials. It means the organisation can push back against concerns about cultural relativism, by delegating them to national chapters.

TI PNG does have some autonomy to interpret and reinterpret corruption within the local context. This has resulted in the organisation presenting different definitions of corruption that sometimes fail to align to the definitions presented by its parent organisation. In a brochure dated 15 November, 2005, TI PNG defined corruption as: 'the abuse by an individual of a public position to acquire illicit personal profit' (Transparency International Papua New Guinea, 2005). Moreover, a brochure marking the organisation's tenth year of operating in the country defined corruption as 'the abuse by a person of a public position to get personal benefits illegally' (Transparency International Papua New Guinea, 2007). The differences between these definitions and that of TI PNG's parent organisation are subtle, yet profound. Both of TI PNG's definitions stress the individual and refer to public office, whereas its parent organisation's broader definition (the abuse of power for private gain) does not (although TI has referred to the public office in their official definitions of corruption in the past and in some research instruments [Holmes, 2006a]). As discussed, internationally the TI 'movement' has ostensively moved away from definitions that focus on the state. In addition, TI PNG's definition also changed from suggesting corruption was 'illicit', a word that alludes to the breaking of both legal and ethical rules or guidelines, to 'illegal', a narrower term concerned with the law.

One way to read these changes is that at one time TI PNG saw corruption more narrowly than their international parent organisation. Another explanation is that the change is arbitrary, reflecting the changing opinions of personnel who compiled this material. Whatever the reasons, this change indicates that the agency has struggled to maintain a consistent definition of the term over time within its official literature. This shows that TI PNG has some flexibility around how corruption might be interpreted within the PNG context.

As we saw in chapter 2, international anti-corruption agencies are mostly concerned with corruption involving state officials. Likewise, in official documents TI PNG highlighted the ways in which state officials – especially politicians and public servants – caused corruption. For example, press releases and media appearances around the time of fieldwork highlighted a range of government decisions it believed hindered good governance, including: the government reneging on an investigation into the finance department (*PNG Post-Courier*, 2008), the purchase of a 'luxury' vehicle for the West New Britain provincial administrator (Rheeney & Sasere, 2002), the government failing to follow bureaucratic procedures (Transparency International Papua New Guinea, 2008b), unnecessary wage increases for MPs (*The National*, 2009) and the proposed purchase of firearms by the Correctional Services Minister (Kelola, 2010: 15). The agency acknowledged that there can be 'a growing misconception that corruption exists only in the public sector' (Transparency International Papua New Guinea, 2008c). However, in line with the mainstream view, most public statements reinforced the primacy of public officials' involvement in corrupt activities and the failure of the government to investigate alleged corruption (see Walton, 2013b).

TI PNG's official narratives about the causes of and solutions to corruption revolved around two issues: the inadequacies of existing state-based institutions and citizens' engagement with the state. The agency has a long history in trying to strengthen or replace state institutions. It was involved in advocating for a change in PNG's electoral system in the 1990s, which ultimately helped to redefine PNG's voting process. For more than a quarter of a century after independence PNG had a first-past-the-post voting system which meant that voters chose only one candidate on the ballot. This method meant the candidate with the highest amount of votes won, even if they received a very low proportion of the overall vote. In PNG this system resulted in politicians often being elected with less than 20 per cent of the tally. In 2001 the Morauta government introduced electoral reforms designed to help address corruption, including introducing a system of Limited Preferential Voting (LPV). This system was first used in the national elections in 2007 (prior to this it was used sporadically in by-elections) and involves a complicated process of excluding candidates with the lowest number of votes and reallocating preferences to remaining candidates until a winner is determined. Through this process winners are eventually allocated a majority of the votes cast. This significant change was helped along by TI PNG and other development actors.

In 2008, TI PNG sought to improve transparency and accountability of the government through its 'legislative initiative' strategy. This involved the organisation contributing to debates on legislative initiatives such as proposals to institute an Independent Commission Against Corruption (ICAC) and an Extractive Industries Transparency Initiative (EITI). Over time the organisation's efforts contributed to the establishment of both of these institutions; the former was signed into the country's Organic Law (a law second only to the constitution) in 2015 and the PNG government signed up to the latter in 2013. So the agency has, at times successfully, helped to reshape the PNG state through advocating for new state institutions to help address corruption.

The second way TI PNG has framed the causes of, and provided responses to, corruption in PNG is in relation to the role of non-state actors. In 2008 TI PNG focused on five strategic priorities aimed at addressing corruption: advocacy, civil and electoral education, memberships and building coalitions, legislative initiatives and policy research. Four of these strategies – advocacy, civil and electoral education, policy research and memberships, and building coalitions – directly related to building the capacity and awareness of civil society. In turn the organisation prioritised activities that sought to empower citizens. Its electoral campaigns and school curriculum programmes exemplify its focus on empowering citizens. In the lead-up to the 2007 elections the organisation provided provincial election awareness seminars; printed teaching kits and posters about the LPV system; trained the public in election observation; and educated eligible voters on the LPV system, principles of good governance and democracy. In the lead-up to the 2012 elections the organisation engaged in similar activities. In 2009 the organisation provided training materials for teachers in schools across the country; these materials laid out lesson plans for teachers to guide student discussion about corruption and good government.

The agency's official documents suggested a potential hardening of attitudes towards the role of culture in fostering corruption. In 2003 TI's Secretariat engaged two academics from the University of Papua New Guinea – Papua New Guineans Albert Mellam (at the time of writing the vice-chancellor of the University of PNG) and Daniel Aloi – to write up a report on the country's National Integrity Systems using a framework devised by the TI Secretariat. The work was supported by the Institute for National Affairs and TI PNG's future Chairman (2003–2008) Mike Manning. As academics, Mellam and Aloi (2003: 12) took a nuanced approach to the links between culture and corruption:

> Culture in PNG is one factor that can give an explanation to the causes of corruption. While it can be argued that there are certain attributes of the culture which seem to be more compatible with corruption, this does not mean that PNG has a corrupt culture. In many parts of the country, sharing and caring is synonymous with a clan leader or an elder of a family. Amongst traditional leaders, mobilisation and distribution of wealth is an essential component of their responsibilities.

A few years later the organisation was painting the relationship between culture and corruption in more dire terms. In the organisation's 2007 Annual Report then Chairman Mike Manning said that fighting corruption would entail 'a change in the attitude of the whole community' (Transparency International Papua New Guinea, 2008a: 4).

The organisation also sought out the private sector as important partners in the fight against corruption. In 2009 it supported the Business Against Corruption Alliance (BACA), which was set up as an initiative of TI PNG to 'ensure businesses do not bribe or offer gifts to government officials to get contracts and permits to work in the country' (United Nations Development Program, 2009). BACA was awarded US$20,000 from the British High Commission in PNG to bring together more than 200 members of the Port

Moresby Chamber of Commerce and Industry (PMCCI) in the fight against corruption in PNG (United Nations Development Program, 2009). Its activities included producing workplace codes of conduct, and providing workshops and education materials. The PMCCI, which handled donations, helped manage BACA. So reflecting the economic perspective of the mainstream view, TI PNG supported the private sector to effectively regulate itself against participating in corrupt activity, which was a different approach to its programmes aimed at empowering citizens to keep the state accountable. Having said this, the organisation sought to introduce Integrity Pacts – which are a tool for preventing corruption in public contracting, with companies bidding for contracts agreeing to abstain from bribery, with actors overseen by civil society groups – in the early 2000s. These pacts failed to materialise for reasons unknown (Larmour, 2003).

Unofficial responses

TI PNG's staff and board members are familiar with the international anti-corruption industry, which, as we have seen, tends to reflect the mainstream view and universalise corruption. However, when I visited the organisation in 2008 and 2009, many I met did not share the same certitude about the meaning of corruption as their international affiliates. Respondents stressed that ordinary Papua New Guineans struggled to define corruption, which oftentimes mean they were more likely to support corruption than resist it. A senior respondent said that most Papua New Guineans had such a limited understanding of the concept that they 'do not even know they are being corrupt'. This respondent suggested that poor understanding of how the state functioned, and its laws and rules inhibited an understanding of corrupt as opposed to non-corrupt behaviour. This sympathetic response was shared by another employee, who more directly challenged the mainstream view. Reflecting the critical perspective, this respondent said that, 'the definition of corruption must come from the villagers, how they perceive it'. Despite their connections to the international anti-corruption industry those associated with TI PNG tended to reflect a more alternative view than the TI Secretariat. This suggests that they were – at least in part – taking their concerns about corruption from national, rather than international, cues.

Uncertainty about the nature of corruption unsettled staff, making some question the organisation's messages. For example, some questioned the meaning of corruption shortly after the organisation distributed a press release condemning Police Commissioner Gari Baki's order to shoot-to-kill 'criminals' wearing stolen police uniforms. Baki made the order in July 2008, soon after local currency, then valued at US$800,000, was stolen from the Bank of South Pacific in Madang Province (Radio New Zealand International, 2008). I asked some employees if they thought it was TI PNG's role to challenge the Police Commissioner's decision. One respondent admitted that it is difficult to know how to translate definitions of corruption into real-life situations and, therefore, what issues TI PNG should respond to. During another conversation a group of respondents wondered whether rape could be a form of corruption and if the organisation should campaign about that issue. The lack of certainty about what constituted corruption was so important that a senior respondent suggested that the organisation should be reframing their messages about corruption:

> We need to define the issue of corruption and link it with stealing, because I think people understand stealing. We need to keep it simple, it's similar [to stealing], [but] people have difficulty understanding.

The disconnect between elite interpretations of corruption and those of citizens shaped TI PNG's activities. If Papua New Guineans didn't know what corruption was, they were more likely to support it, and less likely to support TI PNG's anti-corruption programmes. Such concerns led to a research programme to get to the bottom of what Papua New Guineans understood by the concept of corruption, which was conducted between 2008 and 2011. It is this research – which I devised and, along with others, helped to manage – that I draw upon in chapter 3 of this book, to highlight Papua New Guinean perspectives on corruption. In supporting this research, it is clear that many – even within senior positions – within the agency were not convinced that Papua New Guineans shared the same concerns about corruption as those in the international anti-corruption industry.

There were also some who reflected more 'hard-line' (universal) definitions of corruption within the organisation. One respondent argued passionately that definitions of corruption were largely irrelevant because it was the law that determined what corruption was or was not. Such responses were more likely to come from senior members within the organisation who had power over shaping official communications. But this dismissive approach was relatively rare; most respondents were either perplexed by the broad nature of corruption or contemplative about how it could be defined within the PNG context.

In PNG, talk about corruption often results in discussions about politics. Around the country millions of kina are spent on election campaigning, with those who get into office rumoured to receive their investment back and more through various forms of corruption. The political system works on relationships of patronage, whereby political appointees are elected in return for providing goods and money to voters. If elected, candidates often direct development assistance to their own constituents and supporter base, to the exclusion of other citizens. Given this, many involved with TI PNG focused on the ways in which patronage politics contributed to corrupt outcomes. For example, one board member emphasised that corruption starts with 'electoral breakdown' and is marked by patronage between leaders and their constituents. Dysfunction in the electoral system was of particular concern for respondents because it was perceived to infect the bureaucracy. Reflecting upon the effect of the electoral system on the public service another board member commented: 'when you consider [the state of the electoral process] it's not surprising that the other elements of government are not working. The lack of accountability starts from the top, and flows right through those organisations.' He went on to explain that due to patron–client relationships between elected leaders and public servants, 'departments are being politicised' which has resulted in 'departments and their processes and systems deteriorating to the point of being inefficient'. The difference between TI PNG's engagement with politics and those of local actors is further explored in chapter 6; but for now it is worthwhile noting that – as these quotes demonstrate – stakeholders' response to the politics of corruption was directed towards building a Weberian state bureaucracy. That is, it reflected a strong public office perspective (a part of the mainstream view), in that respondents hoped to segregate politics from patronage and politics from the bureaucracy.

In turn, respondents framed the organisation's activities in terms of bringing about Weberian norms in a variety of circumstances. Some focused on the need to untangle the entwinement of voter and candidate relations. One respondent said ongoing voter education programmes were required because only continual education could break the 'cycle of dependency' built up between voters and politicians. Senior respondents saw this process as a part of a broader project that sought to develop an impartial Weberian bureaucracy free from political interference. One senior respondent explained:

> The work of TI and other groups is to try and strengthen the electoral process first and then, as a result of that, start looking at public sector reform; widespread public sector reform, so we can isolate public servants from widespread political influence.

Given this framing, those who were interviewed – both senior and junior stakeholders – often stressed that culture was a significant pathway to corruption. Indeed, respondents were concerned that corruption was occurring in PNG because citizens' 'traditional' values were failing to adapt to the expectations of modern systems of government. For one employee, the ubiquity of the *wantok* system meant that Papua New Guineans had a 'misconception or misinterpretation of how far personal obligations should go'. In other words, citizens had yet to redirect their 'traditional' alliances to family and clan, to displaying an allegiance to the state. For some, this was tied to Papua New Guineans' lack of understanding about Western ethics. Relating to me as a fellow Westerner, one board member said that before colonisation:

> I do not think there was traditional ethics in terms of what you and I would describe as ethics. I do not think it's a concept. I think that's why there is this difficulty [with corruption].

For this respondent Papua New Guinean traditional values were opposed to Western ethics. Indeed, he went on to stress (reflecting Giddens' [1984] argument) that traditional societies feature a lack of free individual association and decision making) that Papua New Guinean society was marked by communities where a few made decisions and the rest followed, a situation markedly different to the individualism of Western societies. This respondent highlighted the broader concern of some associated with TI PNG: that traditional values were not adequate for responding to the threat of corruption, and that they jarred with the expectations, rules, regulations and laws of modern institutions. This view was somewhat similar to calls within the organisation's annual report for citizens to change their attitudes towards corrupt practices (Transparency International Papua New Guinea, 2008a).

This Weberian vision – which echoed the public office perspective – was reflected in the organisation's programmes. In 2009, the organisation provided training materials for teachers in schools across the country; these materials laid out lesson plans for teachers to guide student discussion about corruption and good government. A senior respondent believed that educating citizens had helped shape local ethics in line with Western norms since the organisation had started operating in the country:

> Since 1997, in terms of trying to change the curriculum at schools, [TI PNG] have [advocated for] ethics education to be emphasised. That's something which would sound natural for someone coming from a Western background, but when you consider most of these people are only the second generation exposed to these Western concepts, certainly no more than third … prior to that there really was not much point in ethics because you did not have choices.

For other respondents enabling this shift in ethics meant targeting the young – which TI PNG did through supporting the Youth Against Corruption group – because young people were, unlike older generations, perceived to be less likely to know the difference between right and wrong. One employee explained, 'our young people here, their minds are, how should I say, they're not broad. They do not know what corruption is, what effects it has.' This employee went on to explain that the youth of PNG were the future leaders of the nation and 'might turn this whole thing of corruption around into something good'. The young were considered a group more likely to be in need of having their ethical understandings changed.

The transnational threat of corruption was mostly associated with the operations of Asian businesses operating in the country. A senior respondent emphatically stated, 'there is no doubt that the growth of illegal Asian business in PNG over the last 10 years has had an impact on corruption.' Respondents were concerned that Asian businesses took advantage of weak government regulations. For example, one board member said:

> We're getting more and more businesses, predominantly of an Asian origin, who, and I do not mean to sound jingoistic about it, [but they] know that the regulatory authority is lax and take advantage of it.

Others expressed similar sentiments, and were similarly careful not to appear racist. Some suggested corruption was due to Asian organisations' management structures, rather than the race of their employees. For example, one respondent cautiously noted:

> Asian businesses are private companies and they're a little bit less accountable [than Western firms], they do not have to face up to annual general meetings and so on; [for example], logging companies, these are all big family businesses and they have strong internal allegiances and obligations … I think it's also largely not so much a cultural thing, it's just the way some of the businesses are organised.

So, for this respondent the involvement of Asians in corruption was related to the way their firms operated, rather than their nationality per se.

It is not surprising that the corruption by Asian businesses was not referred to specifically in public documents; anti-Asian sentiment is an explosive issue in PNG. As discussed in the previous chapter this sentiment has been exploited by a group of local anti-corruption activists – the Coalition – who have used it to help create havoc across the country. What is telling about this response is that it presented a very particular interpretation of transnational corruption in PNG, one where Asian business practices were suspect while those of Western organisations were by and large a model of good governance – which reflects the mainstream view's economic perspective. This is not to say that respondents were always at ease with the practices of Western transnational companies. Some expressed concern about TI PNG receiving funding from mining operations, Ok Tedi and Lihir for example. However, these concerns were expressed more in terms of the environmental damage caused by such mining companies rather than the part they played in corruption per se.

TI PNG's narratives on corruption then were more ambivalent than those of their parent agency TI. Official accounts of corruption were often in line with the broader mainstream approach championed by its parent organisation. (Although, given its degree of autonomy, there were a number of instances where even official accounts of corruption challenged international discourse.) The unofficial accounts provided clearer challenges to the mainstream view; however, respondents clearly saw their role as changing Papua New Guinean ethics to reflect the norms of Western institutions. While there were concerns about the mainstream view, ultimately respondents believed that this view would prevail – and PNG's citizens needed to adapt.

The Australian aid programme

The Australian aid programme was known as the Australian Agency for International Development (AusAID) during the time of my fieldwork. The agency traced its origins all the way back to 1974, when it was known as the Australian Development Assistance Agency. Since that time it had gone through a series of name changes before becoming known as AusAID in 1995 (AusAID, n.d.). However, with the election of a coalition government in 2013 – led by the conservative Liberal party – the agency failed to reach its fortieth birthday. In early 2014 it was integrated into the Department of Foreign Affairs and Trade (DFAT) – which meant Australia's diplomatic and aid programmes were forcibly stuck together – and Australia no longer had a stand-alone aid agency. This section takes a look back at the operations of an independent aid programme that has, in recent years, diminished in size and importance, while expanding its focus on corruption.

The Australian aid programme works mostly through partner governments. In the late 2000s this approach was guided by the *Port Moresby Declaration*, in which Australia declared its commitment to bringing about a 'new era of cooperation' with Pacific nations while respecting the independence of these nations (Commonwealth of Australia, 2008). In PNG, the aid agency supported the Australia–Papua New Guinea Partnership for Development agreement (The Government of Australia & The Government of Papua New Guinea, 2008), which was signed by both governments. The agreement committed each government to advance the partnership by 'reinforcing Papua New Guinea's leadership and ensuring good international practice in development approaches' (The Government of Australia & The Government of Papua New Guinea, 2008: 2). While Australia significantly shapes the nature of these agreements, they play a role in orienting the goals and activities of the aid programme towards those of recipient governments.

While corruption has been high on the Australian aid programme's agenda for some time, it wasn't always the case. The importance of governance – often a euphemism for anti-corruption – within the AusAID programme increased after the 'Simons Report' (Warr, 1997) of 1997, with the government's response to this document, 'Better Aid' (Downer, 1997), indicating that, for the first time, it would make governance a specific focus of the Australian aid programme. From that time the prioritisation of good governance policy grew quickly, until almost a decade later the (still Liberal-led coalition) government released its first stand-alone anti-corruption policy.

This anti-corruption policy was launched in March 2007 (AusAID, 2007c) – the year I made a short trip to PNG to assess logistics and potential respondents for my research. The policy reflected the then coalition government's commitment to openly addressing corruption. It set out three 'mutually reinforcing elements' to its strategy: building constituencies for anti-corruption reform, reducing opportunities for corruption and changing incentives for corrupt behaviour (AusAID, 2007c: 1). The policy set aside specific resources for anti-corruption activity and foreshadowed a greater openness around AusAID's anti-corruption agenda. While previous Australian governments had mostly tiptoed around corruption, that government – led by Prime Minister John Howard – made it a priority.

The amount of emphasis that the Australian aid programme placed on corruption changed over time. The Labor government led by PM Kevin Rudd

(PM from 2007–2010 and in 2013) and PM Julie Gillard (2010–2013) downplayed 'corruption', focusing instead on achieving the Millennium Development Goals – which did not include explicit goals and targets on corruption (unlike the more recently agreed Sustainable Development Goals). This is evident through an examination of two annual reports released under the Liberal-led coalition and Labor governments. AusAID's 2006–07 annual report, released under the Howard Government, nominated anti-corruption as central to its development strategy; the word 'corruption' featured 32 times in the 267-page document (AusAID, 2007a). The agency's 2008–09 annual report, released under the Labor government, focused less on addressing corruption and more on meeting the targets of the Millennium Development Goals (AusAID, 2009a); the word 'corruption' featured 16 times in 330 pages. Still, since 2007 anti-corruption has continued to be a key concern of the aid programme, which is highlighted by an analysis of the way the programme has spent its funding.

During PM John Howard's reign (1996–2007), the percentage of Australia's overall foreign aid budget directed to governance programmes increased dramatically – from just under 5 per cent in 1996–97 (Cirillo, 2006: 47), to peak at 31 per cent in 2005–06 (AusAID, 2006a: 6). In line with their emphasis on the Millennium Development Goals, the election of the Labor Government in December 2007 led to a reduction in overall funding on governance and anti-corruption programmes. In the 2009–10 budget 22 per cent of all aid was directed towards governance-related programmes (Smith & McMullan, 2009). However, in PNG, governance initiatives continued to be prioritised.

In the 2008–09 financial year AusAID's country programme aid to PNG totalled AU\$355.9 million (US\$263 million), while assistance through other government departments and AusAID's regional and bilateral programmes brought the total of aid to PNG to AU\$400.3 million (US\$296 million) (AusAID, 2009a: 36). The proportion of Australian aid spent on governance activities in PNG actually increased under the Labor government. In the 2006–07 period, 33 per cent of funding was spent on governance in PNG (AusAID, 2007a). Between 2008 and 2009 an estimated 49 per cent of Australian official development assistance (ODA) was directed towards improving governance, far outstripping funding in other areas such as health (18 per cent) and education (12 per cent) (AusAID, 2009a: 37). The Labor government thus downplayed talk about corruption and reduced financing for governance programmes in the overall aid programme, but increased the allocation of aid spent on governance in PNG.

In 2007–08 PNG received funding through AusAID's 'Anti-Corruption for Development' initiative, which aimed to increase demand for anti-corruption reforms (AusAID, 2007b). Funding was allocated to develop an anti-corruption plan, and strengthen auditing procedures and other oversight institutions and accountability systems. Funding was also made available for 'research on the links between governance, corruption and resource industries' (AusAID, 2007b: 1). After the election of the Labor government in 2007 anti-corruption initiatives continued to be funded through the agency's governance programme. Through this support AusAID funded and provided technical assistance to the National Anti-Corruption Alliance (a coalition that has been responsible for investigating corrupt activity), NGOs such as TI PNG, the OC PNG, and initiatives that sought to improve the financial management and capacity of PNG provincial and local governments (AusAID, 2009a). While the governance programme incorporated a wide variety of activities, many of these were designed to have a direct or indirect impact on corruption (AusAID, 2007a, 2008, 2009a, 2009b, 2010).

The heart of the old aid programme (as well as its recent incarnation under DFAT) was in the Australian High Commission located only a few hundred metres from PNG's Parliament House, in Waigani – a suburb of Port Moresby, which is home to the PNG Parliament, an array of government departments and alongside foreign embassies. As one might expect from a foreign outpost in one of the most dangerous cities in the world, the High Commission was ringed by tall fencing and patrolled by serious-looking security guards. Getting into the place requires walking through a metal detector, signing in, and providing details to a receptionist behind what looks like bulletproof glass. Concerns over security make this an opaque space, open only for those who are signed in and can procure the appropriate lanyard. While the High Commission is at its heart, AusAIDers and development contractors were placed around the city (mostly) and regional centres, mostly as contractors. In 2009 the agency officially employed 23 people in PNG (AusAID, 2009a: 285), but through development contractors it effectively engaged many more.

Official views on corruption

Officially, AusAID defined corruption broadly. In the organisation's anti-corruption policy, corruption was defined as 'the misuse of entrusted power for private gain' (AusAID, 2007c: 3), the same definition as that given by TI (Transparency International, n.d.-a). In this document AusAID also acknowledged that corruption 'can take many forms and vary depending on local culture and context' (AusAID, 2007c: 3). Accordingly, the organisation stated that corruption can include petty corruption (such as bribes or illicit payments for routine bureaucratic processes) through to state capture (collusion between the private sector and politicians or public officials for their own private, or mutual benefit). AusAID noted the latter can lead to the government introducing laws that unfairly favour vested interests. The agency's broad and inclusive definition suggests that corruption may occur in many situations and take different forms.

AusAID's official documents cited a range of reasons for corruption occurring. At the international level, its inaugural anti-corruption policy (AusAID, 2007c) – a policy aimed at guiding anti-corruption activities across various country programmes – attributes corruption to issues such as weak or non-existent public sector rules and regulations, political appointments, low wages and ineffective oversight mechanisms. The Papua New Guinea–Australia Development Cooperation Strategy 2006–2010 asserted that the absence of 'improved governance and public administration' caused corruption (AusAID, 2007c: 13). The Office of Development Effectiveness (ODE) – a unit that was separated from AusAID's programme management – while cautious, was more explicit. It linked corruption to weak institutions, poor rule of law, crime, poor enforcement of property rights and poor quality of public administration (Office of Development Effectiveness, 2007: 19). So, the state was central to the way the organisation officially framed the causes of corruption.

In the primary period of research (2008–09), AusAID most prominently employed two approaches to mitigate possible corruption. First, the agency addressed corruption through technical assistance (TA)[1] programmes. TA – estimated to account for half of Australia's total contribution to PNG in the late 2000s (Independent Review Team, 2010) – involved, in part, the placement of Australian experts in key positions within the PNG government to transfer capacity and knowledge, and monitor systems of governance (AusAID, 2008). Programmes included under TA included the Ombudsman Commission Institutional Strengthening Project (OCISP), the PNG–Australia Treasury Twinning Scheme (PATTS) and the PNG Advisory Support Facility (ASF) (Phase I) (Ives, Midire, & Heijkoop, 2004). Each of these programmes featured Australian consultants working with Papua New Guinean partners. Consultants assisted the implementation of a range of 'practical targets' (including planning, monitoring and expenditure controls) that sought to improve the effectiveness and efficiency of Papua New Guinean state departments and organisations (Ives et al., 2004).

While each of these programmes met with varied success (a review of these three projects showed that while the OCISP was a success story, the PATTS and ASF showed mixed results 'at best' [Ives et al., 2004: x]), some official documents claim that TA has been effective in addressing corruption. For example, according to the Office of Development Effectiveness (2007: ii), Australian TA helped to shut down over 450 trust accounts that were being used to circumvent expenditure controls. Despite criticisms – including by the then PM Michael Somare (Keane, 2010) – and calls for TA to be scaled back (Independent Review Team, 2010), it was, and continues to be, regarded as an effective strategy against corruption by the aid programme.

The second approach the agency took to fighting corruption involved supporting non-state actors in their efforts to monitor the government and push for greater transparency and accountability. For example, the Democratic Governance Programme (DGP), a programme that supported private sector and civil society initiatives, stressed that 'anti-corruption initiatives have a high priority' (AusAID, 2006b: 1). The programme sought to fund a range of organisations, including TI PNG and the Media Council of PNG, to demand good governance from the government of PNG (AusAID, 2006b).

The private sector was also officially considered a key partner in the fight against corruption; indeed, the agency underplayed the private sector's tendency to engage in corruption, painting it as an 'anti-corruption champion'. In its touchstone anti-corruption policy paper, 'Tackling Corruption for Growth and Development: A Policy for Australian Development Assistance on Anti-Corruption', whilst acknowledging that the private sector can be a party to corruption, the problem of corruption in the private sector is explained away:

> While foreign companies may be targets for illicit rent seeking behaviour, bribes may also appear necessary to win business. However greater transparency and international competition, and an increased awareness of the costs that corruption imposes on business, have seen the emergence of private sector champions of transparency.
>
> (AusAID, 2007c: 5)

This framing of the private sector as a key partner in the fight against corruption was in line with stakeholders' views about how to best respond to corruption in PNG, as outlined next.

Unofficial views on corruption

When asked about corruption in PNG, respondents reflected the mainstream and alternative views in different ways. Those keen to reflect the agency's official responses to corruption – which was particularly the case with those in more senior positions – reflected the mainstream view. For example one senior adviser reflected the legal perspective, defining corruption as going against 'whatever is in the law'. Another senior PNG-based employee said corruption means:

> To gain personally from a transaction, which is illegal. [It] does not need to be financial. [It can include] any type of personal gain that is illegal.

Like the agency's official attempts at defining corruption, these definitions clearly focused on the state.

More senior respondents focused on corruption caused by public sector employees failing to adhere to government procedures. For example, one employee decried the poor quality of financial management around the country:

> There is such poor record management in the PNG public service. Both the financial records and records of decision making have a habit of being lost or never being recorded here … No one knows what is happening.

In such an environment, this respondent lamented, corruption was bound to occur. Another senior respondent said that, 'one of the problems about corruption is that our [governments] are not managed well'. For these respondents corruption was caused by the inability of public servants to comply with, and enforce, bureaucratic rules and procedures.

This lack of procedural compliance was, it was sometimes claimed, rooted in deficient managerial competence. For a senior respondent, the public service in PNG was not adequately versed in private sector management principles and techniques. Compared to public sector organisations, businesses 'have good leadership that understands the process of an organisation; a private organisation is run well', he said. Others were concerned that the independence of the bureaucracy was undermined by political affiliations. One senior employee said, 'everything seems to be done for political reasons', resulting in a public sector beholden to powerful patrons, rather than acting in the best interest of citizens. These responses highlighted respondents' fear that public officials were not able to follow procedures competently, freely and impartially.

Other respondents argued that corruption was caused by poor knowledge, and enforcement, of the law by those in positions of power. One senior employee said that while there was 'no excuse for breaking the law', many political leaders were 'ignorant of the law'. The problem of corruption was also believed to be compounded by a lack of effective law enforcement. 'There is no enforcement or follow up [on legal cases]. Everyone knows this [and] everyone is unhappy about it,' one advisor lamented.

Many were also concerned that 'traditional' and 'cultural' values helped to perpetuate corruption. For example, one mid-level employee believed 'the attempt to blend the traditional law with the Westminster system has meant people have come to expect corruption'. When politicians come to power, he said, they believed that: 'I am allowed to put my hands in the moneybox because I am the head of the organisation; like I can kill two or three of your pigs and get away with things [as was the case in "traditional" times and is still the case in many Papua New Guinean communities]'. This response highlighted a broader concern within AusAID: that Papua New Guinean traditions had failed to adapt to Western ethics, rules and laws – a view absent from official documents.

It was thus not surprising to hear respondents defending the agency's efforts to instil more Weberian values within the public service. Respondents, for instance, defended the agency's approach to TA. One advisor said, 'the reason that you have international advisors going in [to the government] is that they bring a skill set that is lacking in the country in a particular area.' He added that advisors transferred these skills to local Papua New Guineans to improve management of the public service. Another senior employee suggested that TA could curb the potential for corruption by 'providing a different picture as to how a public service could operate, what a motivated workforce could look like, and an example of good management'. Most argued that TA was crucial to build public servants' capacities and knowledge about how to bring about good bureaucratic practice, and in turn address corruption.

While a review of the PNG–Australia development partnership – which came out after the term of fieldwork – called for advisors to take on a greater anti-corruption role (Independent Review Team, 2010), some respondents highlighted the challenges advisors faced in monitoring corrupt behaviour. One anti-corruption advisor said one of his colleagues 'was getting close to corruption and the PNG government complained about him; [as a consequence] he had to go back to Australia'. As a result, in order to ensure a cordial relationship with their counterparts, advisors spoke of deliberately downplaying their anti-corruption role. One respondent, involved in improving public sector governance, suggested:

> It is very important we do not call it [our work] anti-corruption, we do not say we are going to come in and fight corruption. What we do say is we want better transparency and accountability, better information in terms of collection, and dissemination [of information] to support evidence-based decision-making.

Thus, while respondents stressed the importance that advisors can play in fighting corruption, they were also wary about potential conflicts that can arise from their role.

Respondents also believed that engaging with civil society could help to shift Papua New Guinean attitudes in line with state values. To do so one advisor spoke of needing to bring about a 'fundamental mind shift' within the country, a shift that he believed NGOs were best positioned to bring about. Others suggested that NGOs could help introduce a 'shame culture', where people were shamed out of engaging in corruption by relatives, friends and clan members. Some argued that because Papua New Guineans did not understand the rules and regulations of the state and what is expected of them, NGOs were essential to take that message to the people. One respondent argued that partnering with NGOs could help educate Papua New Guineans in the ways of 'processes and procedures' that are central to a modern state.

Respondents understood that the private sector could be responsible for corruption, yet many stressed that the private sector's role was more about solving corruption than causing it. One said, 'the private sector in Papua New Guinea [is] a very strong force in driving social, economic and other agendas' and thus, he continued, can help the fight against corruption by 'speaking out against it'. The private sector was also lauded for its efficient management practices that, it was argued, helped to decrease corruption in its own ranks. 'Businesses are doing much, much better than the public service [in addressing corruption] because of their good management,' one senior respondent said. For many respondents the public sector could learn from the private sector's management principles. For example, one senior employee said, 'the private sector lights the way [for the government]'. Ultimately, those within the organisation reflected the official creed: that the gaze of accountability should be firmly fixed on the PNG state through engaging the private sector and civil society. This belief was in line with its anti-corruption policy (AusAID, 2007c) and reflected the mainstream view's economic perspective.

While most people I spoke to were cheerleaders for the organisation's official response to corruption, there were a few who were sceptical. Some admitted they were concerned about using the term 'corruption' in PNG given the country's high rates of poverty and the complexities of culture. When asked to define corruption, one staff member based in Canberra suggested that corruption was 'just a word', and that it could not accurately portray the meaning of right and wrong, particularly in a country like PNG where cultural, economic and social pressures were substantially different to those faced by Australians. Similarly, another Canberra-based employee articulated concern about the power relations embedded within labels of 'corruption', suggesting that, 'we [Australians] overdo the whole corruption thing. People go from their own perceptions of right or wrong … it [corruption] has such a connotation of evil.' In so doing, both respondents reflected the critical perspective on corruption.

These more critical respondents suggested that the causes for corruption in PNG could be found by examining the mismatch between Western and Papua New Guinean cultural values. For a few, Western values had the potential to lead people to corruption. One Papua New Guinean respondent said:

> There are Papua New Guinean ways and values that people continue to hold at home in the village … we've gone away from that. Here in a town society … values are being eroded because my kids now sit in front of a TV … we've been conditioned [to be corrupt].

These more critical voices were, however, less certain about how to turn these critiques into policy. Some suggested that the agency needed to pay more atten-tion to cultural values and institutions, such as the *wantok* system. Others spoke of the importance of hybridised approaches to development that were sensitive to both modern and cultural institutions. However, such suggestions – that reflected Papua New Guinean Ways – were mostly rather vague. The most tangible response to these concerns was directing funding into research on local interpretations of corruption (see chapter 3). Those who took an alternative view towards corruption understood the problems of anti-corruption discourse but did not know what could be done to significantly rectify them.

In sum, the Australian aid programme's narratives on corruption mostly reflected the mainstream view of corruption. This was evident from the agency's anti-corruption policies, programmes, and from a number involved with the agency. Although most respondents reflected the mainstream view, a few were aware of the limitations of this within the context of PNG. However, there was little indication that the agency was able or willing to institutiona-lise these conceptual concerns in their responses to corruption and poor governance.

Discussion and conclusion

The international anti-corruption industry has helped to shape narratives about, and responses to, corruption in PNG since the late 1990s, a time when corruption had become a central concern for development agencies around the world. In PNG, the two most dominant anti-corruption organisations – TI PNG and the Australian aid programme – reflected the international anti-corruption industry's mainstream view in different ways. TI PNG saw itself as a more transnational organisation, gaining its legitimacy from PNG's citizens while carrying out a range of activities in line with its international mandate. Even though the agency originates from outside of PNG, TI PNG ostensibly heeds the call from PNG's founding leaders for a hybrid approach to development, one that brings together the strengths of PNG and Western institutions (see chapter 4). This approach, which is a hallmark of TI chapters around the world, has been lauded by Bukovansky (2006) as providing a more meaningful approach in framing and responding to corruption.

What TI PNG's official documents mask, however, are the uncertainties, ambiguities and concerns that those within the organisation have about the nature of corruption. Some within the organisation reflected the concerns of the country's CPC: they wanted to see a more nuanced approach to understanding and responding to corruption, in line with PNG's many cultures. They saw cultural principles as a strength to be drawn upon in the fight against corruption. They knew that some Papua New Guineans were not in a position to resist corruption, and that many did not really know what it meant. The difficulties in translating the alternative view into policy meant that, ultimately, these concerns gave way to responses that sought to shape Papua New Guinean ethics in line with the laws and rules of the state – thus reflecting the mainstream view's public office perspective. The CPC was concerned about corruption involving transnational Western companies; so too were citizens who lived in areas dominated by mining or logging companies (see chapter 3). However, TI PNG – in line with broader international approaches to anti-corruption – reflected an economic perspective by considering the Western private sector as an ally rather than a cause of corruption. So while both alternative and mainstream views are a part of debates about corruption within the organisation, it is the mainstream view that has been most successfully transferred to TI PNG – it most strongly influences both discourse and practice.

The Australian aid programme worked more directly with the PNG state, gaining its legitimacy from the PNG government. This presented a challenge for its anti-corruption initiatives, with those engaged with TA treading carefully so as not to attract the government's ire. Within the organisation there were misgivings about the aid programme's approach to anti-corruption – particularly in regards to cultural issues. Many within the organisation understood that working on corruption in PNG was a messy business, one that demanded attention to the cultural, economic and political issues shaping corruption.

Still, most of the aid programme's approach was based on a technical response to corruption through strengthening and increasing the accountability of Western institutions.

Ultimately, both organisations sought to separate culture and modernity, politics from bureaucracy, and Western and traditional institutions. These anti-corruption efforts were about working towards a type of institutional purity, which PNG culture and politics should ultimately adapt to. The international anti-corruption industry's influence in PNG therefore has helped to reinforce the importance of Western state-based institutions for development. In doing so, the hybrid approaches to development championed by PNG's founding leaders have been, by and large, forgotten.

Note

1 The OECD DAC defines technical assistance (cooperation) as: 'both (a) grants to nationals of aid recipient countries receiving education or training at home or abroad, and (b) payments to consultants, advisers and similar personnel as well as teachers and administrators serving in recipient countries (including the cost of associated equipment)' (Development Co-operation Directorate, n.d.).

6 The (anti-)politics of anti-corruption

Introduction

Chapter 3 highlighted the numerous ways that corruption in PNG – as elsewhere – is linked to the political elite at the local, national and international scales. Given that politicians are never far away from accusations of corruption, it is important to ask how well international and national anti-corruption organisations are positioned to respond to political corruption. Are national and international efforts up to the task of withstanding the pressure that comes with responding to political corruption? And how do their discourses and programmes help or hinder their attempts (if any) to 'speak truth to power'?

Investigating these questions is relevant for understanding how mainstream and alternative views are reflected by these organisations. The mainstream view lends itself to viewing corruption through a technical lens. Get the laws, rules, incentives and the size of the state right, and corruption will dissipate. But politics turns these assumptions on their head, in part because politicians are so often beyond the reach of laws and anti-corruption efforts. Indeed they can write laws themselves or weaken anti-corruption institutions, taking themselves out of harm's way. Politicians can and do intimidate those who seek to expose their wrongdoing. This means an important part of anti-corruption activities, particularly in weak states, is deciding on how they engage with politicians who are accused of corruption (even though not all are corrupt). Given the technical approaches embedded in the mainstream view, political corruption has the potential to disrupt clean mainstream narratives about corruption.

The broader question this chapter examines therefore is: how do responses to political corruption reflect the mainstream or alternative views?

This chapter focuses on the way the Australian aid programme, TI PNG, the Coalition and the OC PNG were shaped by and responded to political corruption in PNG. It starts by examining the ways international agencies have engaged with politics, showing that while they often stick to their technocratic and apolitical principles, they can occasionally challenge political power in ways that run counter to official policy statements. The chapter then focuses on local agencies – the OC PNG and the Coalition – who are far more connected to and shaped by national politics than their international counterparts. These connections – or lack of them – significantly shape responses to political corruption. For local actors, political power and corruption are often critiqued strategically when the threat of retribution is mitigated, or when critiquing the state provides a path to political office. The chapter concludes by highlighting how these responses reflect the mainstream and alternative views on corruption, and what this means for the effectiveness of responses to political corruption in PNG.

International responses to political corruption

International anti-corruption agencies have a difficult time in engaging with politics, in part because their ability to work within developing countries is dependent upon the support of partner governments. Tread on too many toes and all of a sudden political support can significantly diminish. This is a conundrum for international agencies as they seek to strike a balance between maintaining a diplomatic relationship with the partner government and tackling corruption – an issue now central to much multilateral, donor and NGO engagement in developing countries. How international agencies manage this tension is central to understanding the limitations and potential of international anti-corruption efforts within developing countries.

Geopolitics and anti-corruption: the timidity of the Australian aid programme

Standing at the podium of Canberra's National Press Club, Julie Bishop was composed; she radiated a congeniality that long-serving politicians perfect over years of plying their craft. The day of 18 June, 2014 was a big moment for the Minister for Foreign Affairs and Deputy Director of Australia's Liberal party; she was about to launch the Australian government's new international development policy. Bishop was sworn in on 18 September, 2013 as the nation's first female Foreign Minister, shortly after the Liberal/National Coalition government swept to power. The launch of what she called 'the New Aid Paradigm' came on the back of unprecedented cuts to the Australian aid programme. The new government had also merged the aid agency, AusAID, with the Department of Foreign Affairs and Trade. The Minister's portfolio had shrunk even before she'd been sworn in, and this launch offered a key opportunity to stamp her authority over a key, yet diminished, part of her portfolio.

The room was filled with journalists, policy wonks, diplomats, business types and academics who fed on grilled salmon. I was also there with colleagues from the Development Policy Centre. However, the more the Minister spoke the harder it was to enjoy the meal. Bishop's explanation for Australia's largest ever aid cuts left some (including me) with indigestion. On a per capita basis Australia, coming off of a mining boom, was one of the richest nations on earth; however, Bishop calmly explained that the country's debt was a key reason for reducing the aid budget. For a long time aid has made up a tiny proportion of Australian budgets (since the mid-1990s aid had made up around one per cent of the federal budget, compared to defence which can be five to six times that); Bishop effectively argued that cutting aid would help put Australia's books back into the black. 'The New Aid Paradigm', as the agency called this new approach, was – in terms of funding allocated – more about responding to domestic concerns (particularly the country's overall budgetary position) than responding to the great development challenges of our time (such as poverty, inequality, climate change and environmental degradation). As is often the case in the world of aid policy, national politics trumped real commitment to addressing the challenges facing poor countries and poor people.

Under the Coalition's New Aid Paradigm anti-corruption was considered a top-ten priority. The anti-corruption target called for 'New fraud and anti-corruption strategies by July, 2015' for partner countries. As Howes and Negin (2014) pointed out shortly after the launch, this new target built on the broader Fraud Control Plan for the entire aid programme along with the more ad hoc country strategies – initiatives introduced by the previous Labor government. However, featuring anti-corruption as a central part of the aid programme signalled that the Coalition was back in power – focusing on anti-corruption certainly differentiated this new framework from their predecessors, who had played it down in official documents (see chapter 5). Recent changes to Australia's aid budget have further highlighted the importance that the current (at time of writing, October 2016) coalition government places on governance and anti-corruption. In May 2016 it was announced that while there would be cuts to the overall aid budget, governance – which includes spending on anti-corruption efforts and border programmes that promote the rule of law – would be spared. The governance sector was protected from AU $220million (US$162 million) worth of cuts in the 2016–17 budget, making it the largest sector for aid spending in the Australian aid programme for the second year running (Howes, 2016).

At the end of Bishop's speech, as is customary at these events, the media were invited to ask questions. By far the most interesting question was asked by Shalailah Medhora from SBS Television: 'How much confidence do you have in the PNG government given allegations of corruption against Peter O'Neill and his decision to sack the man who brought forward those allegations [Sam Koim, head of Investigation Taskforce Sweep (ITFS)]?' (Bishop, 2014). Bishop paused for but a nanosecond before answering that, 'PNG is one of our dearest, closest friends. Papua New Guineans are family' (Bishop, 2014) and that Australia would support the country no matter what. Although she did say that she had had some frank discussions with her PNG counterparts (although not about corruption per se):

> I've made it clear to the PNG government, and they agree, that we can't continue spending billions and billions in PNG when they won't meet any

of their Millennium Development Goals by 2015, indeed they're going backwards.

(Bishop, 2014)

Shalailah Medhora's question was timely given the recent machinations surrounding PNG PM Peter O'Neill. The week before Bishop's speech in June 2014, Investigation Taskforce Sweep moved on O'Neill over accusations that he had signed away 72 million kina (US$23 million) of the state's largesse for fraudulent legal fees to a firm headed by Paul Paraka, who was himself facing criminal charges from ITFS investigations. An arrest warrant was issued, which O'Neill refused to recognise. To date he has managed to stay in power by installing a Police Commissioner sympathetic to his cause (see chapter 3). At the same time he defunded ITFS, severely weakening its ability to operate. O'Neill also sacked others sympathetic to ITFS' cause, including the country's attorney general and the deputy police commissioner, and suspended the assistant commissioner.

While not mentioned at the time, Medhora's question highlighted the disjuncture between the New Aid Paradigm's focus on anti-corruption and the national and international realpolitik that shapes corruption and responses to it. Ostensibly Australia's new approach to anti-corruption is framed by these political machinations through a better understanding of the local context. To promote effective governance, the new Development Assistance Budget (2014–2015) – which accompanied the policy – committed the government to tailoring 'support to the political context', and conducting political economy analysis to help address 'challenges in fragile and conflict affected situations [a description that fits PNG well], including unequal access to the benefits of economic growth and employment, political alienation and a sense of injustice, which can lead to conflict' (Department of Foreign Affairs and Trade, 2014a: 12). Along with this political analysis, the government called for new anti-corruption policies at the country and regional levels by mid-2015. According to the policy, these plans:

> will detail the measures we [Australia] will adopt to protect Australian Government aid funds and how [Australia] will support our partner countries. Support may include public financial management reform programs, funding of civil society organisations that champion anti-corruption and funding other anti-corruption bodies.
>
> (Department of Foreign Affairs and Trade, 2014b: 29)

This analysis and support should be welcomed, but Australia's feeble response to ITFS' effective closure and O'Neill's arrest warrant highlights how difficult it is to implement. And there was something else wrong with this new approach. According to an aid official I spoke to in late 2015, the plans which were touted as the centrepiece of the approach would not be available to the public, which is of course ironic given that they seek to promote accountability and transparency. So shortly after its launch it was becoming more and more apparent that the policy would disappoint many concerned with good governance, accountability and transparency – within both recipient countries and the aid programme.

There have been some signs that Australia is being forced to take a harder line on political corruption in PNG. For a start, there appears to have been a greater willingness from Australia to cooperate with PNG anti-corruption investigators. After ruffling Australian feathers by calling the country the 'Cayman Islands of the Pacific', due to its lax investigatory efforts into money laundering, in a 2012 speech, Sam Koim said that Australian cooperation with his agency increased – despite years of inactivity (Walton, 2013a). At the same time, Australia said little publicly about political corruption in PNG until Global Witness, an international NGO, caught PNG lawyers explaining how corrupt funds were laundered through Australia. The story – broken by SBS, an Australian television station, and the Fairfax newspaper group – and subsequent public concern helped force Australian Foreign Minister Julie Bishop to respond:

> We take very seriously allegations that the proceeds of crime in PNG and the proceeds of corruption can be laundered in Australia … We are working closely with PNG to ensure Australia is not a safe haven for the proceeds of corruption.

Both of Australia's responses – its increased cooperation with PNG's ITFS and Bishop's comments – highlight the power of external actors for shaping anti-corruption policy. Australia was moved by the threat to its domestic and international reputation through these acts, not by the actual nature of corruption in PNG – corruption by elite politicians – which has been considered a serious threat to the country for decades. This suggests that policy responses to elite *political* corruption are driven not by stated foreign policy concerns, but by political leverage and third parties.

Despite Australia's response, there have been no arrests, sanctions or public investigations or inquiries into the matter. No funds have been repatriated from Australia to PNG. While there appears to be increased cooperation behind the scenes, O'Neill's resistance to his arrest warrant, and subsequent defunding of ITFS, resulted in very little meaningful response from Australia.

Australia's reluctance to speak out and substantively act against political corruption – in the absence of third-party pressure – is due to its reduced leverage in PNG. PNG's gross domestic product (GDP) grew by around 12 per cent in 2015. Even though this represents a significant downgrading from earlier estimates over 20 per cent, it still means that PNG was one of the faster-growing economies in the world. In 2016, GDP grew by only 2 per cent; however, with strong longer-term revenue growth due to a resource boom, PNG no longer relies on Australian aid as much as it once did. PNG is working towards an exit strategy for Australian aid (Independent

Review Team, 2010), with some of the country's elites arguing that one of the key goals of economic expansion is to become more independent from Australia (Fauziah, 2009; Dorney, 2016).

In the lead-up to an Australian federal election in 2013, Australia's Labor government, led by PM Kevin Rudd, made a deal with O'Neill that PNG would host Australian-bound refugees in the remote island province of Manus. This deal, which continued under Australia's coalition government, significantly changed the relationship between the two countries. In particular, the agreement meant that Australia became more indebted to PNG politicians than ever. The deal has given PNG strong leverage over Australia given that the influx of refugees into Australia is a hot domestic political issue. Australians vote for political parties that will stem the flow of refugee boats into the country. The Manus detention centre has become a key part of Australia's inhumane, yet politically popular, immigration policy. Australian officials have been concerned that, if PNG were to pull its support, the Australian government would likely face a domestic backlash from voters.

PNG knows it has good leverage over its ex-colonial master. This is exemplified by PM O'Neill's declaration that he would ban foreign advisors as of January 1, 2016 (ABC News, 2015). The threat was partially carried out, with 15 advisory positions in the departments of finance, transport, treasury and justice ending on 31 December, 2015, while 18 advisory positions remained (ABC News, 2016). O'Neill wanted to see advisors recruited by the PNG government instead, a move which seems to have taken the Australian aid programme by surprise. Explaining the rationale behind the decision, O'Neill took aim at Australia specifically:

> As a developing country we don't want handouts, we don't want Australian taxpayer money wasted and we don't want boomerang aid [money from the aid programme that ends up back in Australia through consultancy fees for Australians and Australian businesses].
>
> (ABC News, 2015)

PNG's leverage has meant that Australia has had to step even more carefully than it has in the past. In 2007 the Australian Liberal Foreign Minister, Alexander Downer, arguably went overboard and harmed Australia's relationship with its Pacific neighbours – including PNG – by openly calling their political leaders corrupt (AAP, 2007). Even a modicum of such bold discussion about corruption in PNG is unimaginable now.

This incident reflects Australia's broader reluctance to meaningfully respond to the transnational movement of corrupt funds. Australia is a party to numerous international instruments aimed at combatting corruption and money laundering, including the United Nations Convention Against Corruption, the United Nations Convention against Transnational Organised Crime, the Organisation for Economic Co-operation and Development (OECD) *Convention on Combating Bribery of Foreign Public Officials in International*

Business Transactions and the Financial Action Task Force's (FATF's) 40 Recommendations. It has passed laws – including the Financial Transaction Reports Act 1988, the Proceeds of Crime Act 2005 and the Anti-Money Laundering and Counter-Terrorism Financing Act 2006 – which impose duties of due diligence when dealing with international transactions (Koim, 2013: 247). Yet questions have been raised over Australia's commitment to upholding these instruments. A 2012 evaluation of Australia's implementation of the OECD's anti-bribery convention found that Australia had only had 'one case that has led to foreign bribery prosecutions since it enacted its foreign bribery offence in 1999' and noted that this was 'of serious concern' (OECD Working Group on Bribery, 2012: 19).

A 2015 mutual assessment of Australia's anti-money laundering measures drew attention to poor coordination between the state and federal responses, particularly in the Queensland property market, where Koim (2013) and others believe much of PNG's corrupt money from political elites ends up:

> while ML [money laundering] of foreign illicit proceeds through real estate is perceived to be a risk for Queensland (Gold Coast), Queensland has no ML convictions for this activity. AFP [Australian Federal Police] indicated that it does not focus on this risk, believing this ML activity relates to State level predicates, whereas the Queensland Crime and Corruption Commission stated it does not focus on this risk as it relates to foreign money and is thus a matter for AFP.
>
> (FATF and APG, 2015: 57)

Australia's response is part of a broader reluctance to respond to money laundering. A recent report primarily authored by the international inter-governmental body tasked with combatting money laundering and terrorist financing, the FATF, found that corruption offences are often not prosecuted as 'Australia does not consider that foreign predicate offences [those generating proceeds of crime that can be laundered] are major predicates for [money laundering] in Australia' (FATF and APG, 2015: 57). Given this, it is unsurprising that there are few convictions for money laundering. Nor is it surprising that despite much to suggest that PNG elites have laundered corrupt funds into Australia, the country has not repatriated any illicit funds back to PNG (Koim, 2013).

This reluctance to engage with the political dimensions of national and transnational corruption matters for those who are willing to defend the rule of law in the country. Today, the PNG's growing middle-class often rallies against alleged corruption and efforts to undermine anti-corruption watchdogs. Australia's lack of meaningful response during the attacks on ITFS, and its inactions over accusations of political corruption, led some of these activists to accuse Australia of hypocrisy. Many in PNG, as blogger Keith Jackson said in 2014, judged the Australian government by its response to these events. Unsurprisingly, few are impressed. Martyn Namorong, another PNG social commentator, tweeted at the time of Sam Koim's sacking: 'next time DFAT wanna talk about good governance in PNG remind them of their silence now' (Walton & Howes, 2014).

In sum, Australia has been seriously constrained in its ability to speak out and act against political corruption in PNG. There are also signs that its position is further weakening. In part this is due to Australia's own policies, which means Australia's best chance of meaningfully responding to political corruption in the country will depend upon its ability to reframe the diplomatic relationship, which has come to be defined by PNG hosting Australia's asylum seekers. Abolishing this policy would mean Australia is better placed to do and say more about corruption. (This move should also be encouraged to reduce the inhumane treatment of refugees.) So despite the aid programme's statements about engaging the political dimensions of corruption, Australia is unlikely to do so without a broader change in its geopolitical approach to PNG and the Pacific.

Having said this, as chapter 5 highlighted, Australia works with and through the PNG state, and as such will always be limited in its ability to respond to political corruption. There's only so far that Australia will be willing to push the issue – lest it be accused of political interference. This is not to say that the organisation does not still 'work politically' to gain support for anti-corruption reform. There are instances where the aid programme has built coalitions – including politicians – to support anti-corruption reform. In addition, the aid programme has other ways of responding to political corruption in the country by supporting local and national organisations to engage in ways that it cannot. Indeed, the aid programme has over the years provided small amounts of technical and funding support for both TI PNG and the OC PNG, although it steered clear of the Coalition. The ability of these two organisations to respond to political corruption in ways that the Australian aid programme can't, or won't, will be explored in what follows.

Of big-men, politics and TI PNG

The launch of the New Aid Paradigm in Canberra's salubrious National Press Club was a world away from the public funeral for one of PNG's most outspoken government critics. In an open-air church I along with hundreds of others listened to dignitaries, politicians, family and friends bid farewell to Michael John Manning (1943–2008), or Mike Manning as he was known. Manning was 65 years old when in 2008 he died of a heart attack while going for an early morning walk near his home in Rabaul, in the north-eastern province of East New Britain. The funeral was held on 26 August, 2008.

Born and raised in Australia, Manning relocated to PNG in its year of independence in 1975. Since that time he had seen first-hand the threat that corruption posed to the state, having worked in the public and private sectors – including the finance ministry. But it was his role as a civil society advocate that brought him the most public attention. In 1997 he became the director of the PNG Institute of National Affairs, a private sector-supported think tank, a role in which he served for eight years. During this time he rose up through TI PNG's ranks, first as a member of its board of directors in 1999 and then as the organisation's second chairman, after the death of its first chairman Sir Anthony Siaguru. While Manning was outspoken against government corruption before his move to TI PNG, his association with the organisation helped amplify his concerns; Manning used that platform to further highlight the seemingly growing problem of corruption in the country.

In 2003 he co-authored a paper with Susan Windybank (Windybank & Manning, 2003) arguing that PNG was on the brink of economic paralysis, government collapse and social despair. At the heart of their concerns was the 'systemic and systematic' corruption within public institutions. The publication angered some MPs in PNG. The then PM, Michael Somare, called for Manning to leave the country. That would have been rather difficult given that Manning had renounced his Australian citizenship, had married into Tolai society (from East New Britain Province) and raised his family in PNG. This and other attacks did not deter him: he continued to speak out against government corruption and ruffle the feathers of political elites.

For example, in April 2008 the Somare government disbanded a Commission of Inquiry into the Finance Department, an inquiry it set up in 2006 to investigate reports of widespread corruption. Manning, incensed by the PM's backflip, responded through the media. In a local daily newspaper, the *PNG Post-Courier*, the then TI PNG director said Somare's decision 'was a terrible blow to the credibility of the Government which talks about a commitment to good governance' (*PNG Post-Courier*, 2008). In the same article TI PNG was quoted as supporting the opposition's call for the PM to resign if he could not provide a good explanation for terminating the inquiry (*PNG Post-Courier*, 2008). After such criticism by TI PNG and others, in late April the PM rescinded his decision.[1] In his role of Chairman of TI PNG Manning, in turn, congratulated the PM for his backflip through the media (Palme, 2008). While critical of the PM's decision to halt the inquiry, Manning was willing to provide a gesture of goodwill towards the PM soon after his change of heart.

At a seminar held by TI PNG on anti-corruption day, 9 December, 2007, at the Hideaway Hotel in Port Moresby, Manning claimed that two-thirds of the country's annual revenue was being stolen by politicians and bureaucrats. Rather than being punished for their crimes, Manning suggested that corrupt politicians were being re-elected. PM Somare hit back against these statements, demanding that TI PNG and Manning 'name names' of these allegedly corrupt politicians and turn the proof of wrongdoing over to the police rather than make 'hasty generalisations'. 'Otherwise,' he said, 'they should stop playing power games under the guise of transparency.' But naming and shaming is beyond TI PNG's mandate, making this request redundant.

Across the world, TI is cautious in its engagement with politics. Most chapters, including the Papua New Guinean chapter, stress that the organisation does not reveal names or denounce individuals or companies

(Transparency International Papua New Guinea, 2009a). Its international website highlights that the organisation is 'politically non-partisan' (Transparency International, n.d.-b). TI's influential Source Book – a guidebook for anti-corruption actors – cautions that:

> Many civil society groups are single-minded in the pursuit of their particular cause and have no interest in balancing their aspirations within the wider public good.
>
> (Pope, 2000b: 131)

In turn, apolitical and non-confrontational organisations are considered legitimate representatives of civil society. TI's Source Book suggests that civil society 'gains its legitimacy from promoting the public interest' (Pope, 2000b: 132), yet it also notes that it will only work with groups that are expressly 'non-partisan and non-confrontational' (Pope, 2000b: 135). This resonates with some of the international literature. For instance, Johnston and Kpundeh (2004: 6) argue that when those in the industry recruit local partners in developing countries, they 'should focus on stakeholders who suffer immediate and tangible costs of corruption and have resources they can mobilise against it', particularly in the early stages of building an anti-corruption coalition. This approach, they argue, reduces the chance of anti-corruption 'movements' becoming politicised and thus less legitimate. Manning's critique of politicians and governments meant that, at times, TI PNG was walking a fine line between criticising those who participate in corruption and betraying their non-partisan mandate. Thus far this has worked relatively well for the organisation. Despite sometimes terse responses by prominent politicians the government has not shut it down. In fact, TI PNG has become a 'legitimate' critic of government policy. It has achieved this through its support from prominent Papua New Guineans, representatives from civil society, the churches and the business community (see Larmour, 2003).

After Manning's death in 2008, TI PNG appointed a new interim chairman, Peter Aitsi. Aitsi served until 2011 when board member Lawrence Stephens was appointed. (Stephens was still serving the organisation at the time of writing.) Under these two chairmen TI PNG has responded to numerous government scandals, facing similar push-back, yet less severely than the late Mike Manning. The relative civility of discussion between TI PNG's new leadership and political elites was exemplified by the political response to TI PNG's analysis of the 2012 elections in the country. TI PNG's election monitoring report (Transparency International Papua New Guinea, 2012b) concluded that the elections were 'seriously flawed'. At the time in PNG, these were fighting words. The 2012 election was held after O'Neill replaced Somare as PM in late 2011. Somare was out of the country and receiving medical treatment in Singapore when O'Neill orchestrated the peaceful, but constitutionally dubious, takeover. The election resulted in O'Neill securing his own seat of Ialibu-Pangia in Southern Highlands Province and stitching together a large coalition of MPs to form a government. The election helped provide a veneer of legitimacy for O'Neill orchestrating Somare's dismissal in 2011.

It is then unsurprising that O'Neill responded so strongly to TI PNG's report. In March 2013, the Prime Minister's Office put out a media release condemning TI PNG's claims, stating that:

> Mr O'Neill said he generally applauded and supported the work of Transparency International infighting [*sic*] and exposing corruption and abuse, but he was very disappointed that its claims regarding the conduct of the elections were an exaggeration.
>
> (Prime Minister's Office, 2013)

He was quoted as saying that the overall conduct of the election was an improvement on those in 2002 and 2007, and that with Australian support, the presence of international observers, a free press and a 'robust political party process the election process produced a fair, and democratic, result' (Prime Minister's Office, 2013). The PM targeted TI PNG even though the Domestic Observation Group – another group of monitors involving researchers from the Australian National University – found that the 2012 elections were 'worse than the 2007 elections both in terms of security as well as fraud and malpractice' (Haley & Zubrinich, 2012).

In response, TI PNG Chairman, Lawrence Stephens, defended the organisation's report, noting that it was:

> Based on observations of 282 Papua New Guineans, and some overseas visitors, we have field reports and photographic evidence backing our conclusions.
>
> (*PNG Post-Courier*, 2013b)

Still, Stephens was keen to ensure that the report did not burn bridges between TI PNG and the PM. He said that TI PNG wanted to collaborate with the government, and that it was not an organisation that wanted to 'throw stones' (*PNG Post-Courier*, 2013b). Looking forward to the next scheduled elections, and exemplifying his conciliatory tone, he said that:

> The 2017 National Elections are an opportunity for Prime Minister O'Neill to deliver the best elections in PNG's short history and we are confident of his commitment.
>
> (*PNG Post-Courier*, 2013b)

TI PNG's responses, which mix critique with conciliation, are understandable given that TI PNG has closely worked with the government to promote its anti-corruption messages. In June 2015 TI PNG was able to sign a Memorandum of Understanding with the National Department of Education to collaborate on delivering education materials for PNG schools (Wenambo,

2015). The programme aimed to build on an ongoing commitment by TI PNG to teach students about human rights, democracy, law, advocacy and good governance. With the partnership taking two years to secure, TI PNG could not afford to put the government totally off-side. This demonstrates the difficult balancing act that TI PNG faces when engaging with the PNG government – on the one hand it seeks to censure decisions that contravene good governance, on the other it looks to join with government to spread its messages.

So in some ways TI PNG is subject to the same types of pressures as the Australian aid programme – its sometime benefactor – in that its commitment to working with and through government shapes its political engagement. Yet, unlike the Australian aid programme, TI PNG has been able to politically engage and ruffle elite feathers along the way. This is in part due to the organisational structure of TI chapters, which allows some autonomy for local members to rise through the ranks and stand against what they see as an abuse of the rule of law by political elites. Still, those associated with TI PNG publicly refrain from 'naming names', in line with its international mandate.

The agency also reached out to other civil society organisations and helped coordinate large-scale responses to challenges to good governance emanating from politicians. In 2005 TI PNG, along with the churches, helped scupper two MP bills aimed at reducing the power of the OC PNG and increasing MP-controlled constituency funds by getting tens of thousands of people to sign a petition, which was presented to PNG's Parliament. While this action only temporarily stopped the rise of these constituent funds (as these funds have grown exponentially between 2013 and 2016 [Howes et al., 2014; Flanagan, 2015]), it highlights the potential for TI PNG to effectively direct community outrage about political decisions that undermine good governance.

While the agency continues to engage with community groups, not everyone has been convinced of their political mettle. After the O'Neill government's decision to defer elections for 12 months and introduce a bill to regulate judicial conduct in March 2012, blogger and social commentator Martyn Namorong questioned TI PNG's delay in publicly announcing their participation in public protests over the issue (Namorong, 2012). He suggested that the organisation was captured by the Australian aid programme, which was at the time supporting TI PNG through the Democratic Governance Program. To be fair to the organisation, Stephens and board member Richard Kassman eventually led the rally along with unions and other civil society groups (*PNG Post-Courier*, 2012b). Leading up to the rally, TI PNG and its affiliated Community Coalition Against Corruption also publicly condemned the Judicial Conduct Bill (Transparency International Papua New Guinea, 2012a).

Still, concerns about TI PNG's agenda linger in other circles. Like Namorong, the Coalition were also critical of TI PNG's cautious approach to public protest. This is exemplified by their response to the annual Sir Anthony Siaguru Walk Against Corruption (WAC), TI PNG's main fundraising programme. First staged in 2007, the event secures financial support for

TI PNG by eliciting contributions from businesses, other NGOs and individuals. In 2008, each team donated 2,000 kina to participate, while individuals donated 200 kina. While the WAC has been an important fundraising event, helping TI PNG secure long-term funding in the face of fickle donors,[2] in the media it was presented as a symbol of ordinary people protesting against corruption. The Governor General was quoted in the *PNG Post-Courier* as saying, 'the walk against corruption in Port Moresby and Kokopo (East New Britain Province) was a demonstration of the people's commitment towards stamping out corruption in high places in PNGs administration' (Cuthbert, 2008). TI PNG also promoted this idea by calling on people to 'crusade for a life free of corruption' in a flyer advertising the event (Transparency International Papua New Guinea, 2008d).

The Coalition encouraged their supporters to attend the 2008 WAC. In May, for instance, one leader asked supporters to march alongside the Coalition and TI PNG at the upcoming WAC to show their opposition to corruption. Many in the crowd cheered in response, with one supporter taking up the microphone to urge others in the crowd to attend. On the day the WAC was held, members of the Coalition arrived late, but ready to engage in what they saw as a protest against government corruption. Upon learning that the event was in fact a fundraising exercise and not open to the general public, one of the Coalition's leaders became furious, arguing that the event exemplified TI PNG's reluctance to address 'real corruption': 'This is an event for the elites,' he said. This leader's concern about the exclusivity of the event was not unjustified. At the time of the 2008 WAC, the minimum wage in PNG was 22.96 kina (US$7) per week for unskilled adult workers (High Commission of the Independent State of Papua New Guinea, 2001) – almost ten times less than the cost for an individual to participate in the WAC. Despite attempts to paint the event otherwise, this was not an event for the *grasruts* (*Tok Pisin* for 'grassroots'). For this leader the WAC reinforced the problematic nature of TI PNG's ambiguous stance towards political corruption: 'TI [PNG] only talks. We need action, real action ... we need to kick out [PM] Somare, and we need real leaders to fight corruption.' The event strengthened the Coalition's belief that only they could speak truth to corrupted political power.

National responses to political corruption

Although TI PNG was far more willing to publicly engage in the rough and tumble of PNG's personal politics, both it and the Australian aid programme have been cautious not to push the PNG government too far. National organisations, the Coalition and the OC PNG, have exhibited a far greater willingness to directly engage with the personal politics of corruption. But doing so brings risks of state retaliation. The ways these two organisations were shaped by and responded to the political pressures that come with combatting corruption head-on are explored next.

The Coalition, anti-corruption 'terrorists'

During one of the first encounters I had with the Coalition, in April 2008, it became apparent that they quite enjoyed their role as radical outsiders. Sitting on a bench in a dusty clearing in the shopping suburb of Boroko, one of their leaders identified the group as patriotic terrorists. In a response designed to shock what they probably perceived as a gullible white researcher, this leader insisted that he'd die for his 'beloved Papua New Guinea'. 'I want to blow myself up', he continued, 'just like Osama [bin Laden].' Clearly the Coalition were more willing to directly confront the state than TI PNG – as outlined in chapter 4, at the time of this interview they had just come out of hiding after the Police Commissioner had threatened them. However, in the early stages of my fieldwork I remember wondering about their ability to live up to their radical sentiments. It wasn't long until I got to find out: a few days later the Coalition began what was to become a long-running campaign against alleged corruption in the Somare government.

By May 2008 the Coalition were taking their concerns to the streets two to three times per week. It was during this time when their paranoia about state monitoring spilled over into violence. At a protest, again in the open-air market of Boroko, one leader's oratory was cut short by shouts of 'NIO, NIO!' ringing out from the crowd. NIO is an acronym for National Intelligence Officer, an employee working for the Papua New Guinean National Intelligence Organisation. A member of the crowd had been caught taking photos of the group and the crowd suspected he was spying on them. They reacted angrily. Punches rained down upon the man as he fled the scene. The crowd destroyed the mobile phone he had been using to take photos. At the time Coalition leaders did little to stop the violence and in subsequent protests they taunted those they imagined might be NIOs within the crowd. During one protest a leader sent out a warning to possible NIOs through a hand-held megaphone:

> I want to warn you that some NIO for the NA [National Alliance – the then PM's political party] are around. We also have our own NIO; we are street NIO, you must be aware of that when you move around. If you want to get me, I am from the Highlands, you face me and we will fight.

The NIOs were possibly just figments of the Coalition's imagination – senior police said neither they, nor other security agents, had the resources to monitor the Coalition's protests. But they had some reason to be concerned. Chapter 4 showed that direct threats had come from the then Police Commissioner which had resulted in the Coalition going into hiding.

The Coalition's concern about possible state monitoring did little to dampen the frequency and size of their protests. By July 2008 they were moving their protests around the city, and in some places attracting up to 200–300 people. They were also becoming louder, announcing their concerns about the government through a PA system borrowed from a local church. It was during this time when the Coalition faced off with local police. One of the leaders of the Coalition, Noel Anjo, was asked to put his microphone down and stop protesting. He yelled back that he would do no such thing. The police said they had strict instructions to clear the group out of the market. Anjo stood, defiantly, continuing his harangue against government corruption. A standoff ensued with the Coalition's leaders and supporters facing off against police.

Eventually the police raised their guns and fired warning shots. The crowd scattered; some cowered, dropping to the ground. Anjo reportedly stood still, defiant. As people ran he was reported to yell above the din of the shooting, 'I am not afraid to die for my country. I will die for my country! I will die!' No response came from the police. There was an uneasy silence. He continued, 'I will not move until the Met Sup arrives. Bring me the Met Sup [an abbreviation of Metropolitan Superintendent, a position in the police force, which is in charge of organising security within the National Capital District].' At the time, Fred Yakasa held this position. The protesters had gained approval from Yakasa to conduct their protest and now this leader wanted to negotiate with him directly. The 'Met Sup' eventually arrived and convinced the Coalition to cease their protest.

This was not the last time the Coalition and their supporters were to experience the threat of state violence. In a well-publicised case, Anjo accused the then PM, Somare, of abducting and beating him in reprisal for his activism. According to Anjo, on 16 February, 2009 he was forcibly taken from Waigani (a suburb in Port Moresby) to Mirigini House, the PM's residence, by plain-clothes police and beaten by both Somare and his wife Veronica. In this extraordinary tale, he accused the PM of asking his kidnappers why they had not shot Anjo and dumped his body. He was later allegedly taken to two local police stations – in the suburbs of Gordons and Boroko – and was eventually charged with spreading false information against the PM through his ongoing campaigns. On 25 May, 2009 he appeared in the Boroko District Court and the case brought against Anjo was struck out due to insufficient evidence. On August 7, 2009 Anjo filed a claim with the Attorney-General's office for K1.5 million in compensation from the State – an amount that has not been forthcoming. Later in December 2010 Anjo made a formal complaint to the OC PNG (Radio New Zealand, 2010).

Somare vehemently denied involvement. In Parliament almost two years after the incident he assured members that he and his wife had not been involved, and he noted that no charges had been laid against him. Somare said, 'Only if the matter is taken up in court will I be obligated to answer to these politically-motivated claims' (Gridneff, 2010). Politicians in PNG (and elsewhere) often claim that criticism from civil society and elsewhere is politically motivated. But in this instance Somare's concerns carried some weight, even if the incident did occur in the way Anjo described.

Anjo revealed his kidnapping in a press conference with then opposition leader Belden Namah (from Vanimo-Green River Open electorate) along with opposition members Sam Basil (Bulolo Open electorate) and MP for Anglim South Waghi Open electorate, Jamie Maxton-Graham. The story provided Namah with political ammunition, and he used it to attack the PM in the media (South Pacific Post, 2010). Anjo continued to openly support Namah and the opposition (with photographs of the pair appearing on social media sites) until Don Polye took over as opposition leader in December 2014.

Anjo and those around him were also politically connected in other ways, although they did not like to highlight it. As discussed in chapter 4, elites and some supporters had alleged that prominent politicians might sponsor the Coalition. The Coalition were aware that these accusations could harm their chances of gaining funding from anti-corruption organisations. For example, while I was researching the Coalition in 2008–09 one leader asked me to help develop a funding proposal to raise money for unemployed youths living in the settlement. When asked whom the funding application should be sent to, he replied, 'someone who does not think we're Powes' [Parkop's] organisation' (Parkop was the Governor of the National Capital District at the time and was linked to the Coalition by prominent politicians and the media [Eroro, 2008]). In 2012, Anjo confirmed public opinion about his political aspirations when he made a run for political office himself. After being ejected from the Coalition amid accusations of bribe-taking (Radio New Zealand International, 2012), Anjo ran for the seat of Moresby North West, which he lost to Michael Malabag, MP from the People's National Congress party.

The ramifications of these political connections shaped the public's engagement with the Coalition and, in particular, Anjo. In 2008–09 those who were turning up to their rallies were suspicious of the Coalition's political motivations. A pair attending one of the Coalition's rallies told me that the organisation was likely a front for MPs looking to take over from Somare. Members of the NGO community also said they had reservation about the Coalition because of their propensity for violence and their political connections. With the Coalition's connections to opposition MPs becoming more apparent over time, the group and individuals associated with it struggled to sustain community support.

After he split from the Coalition in 2012, Anjo continued to protest, this time against the O'Neill government. He helped channel public rage at the PM's decision to defund ITFS and dodge an arrest warrant for corruption. Anjo has been a focal point for urban-based protest, gathering small crowds of supporters to protest various corruption scandals. But the public have continued to be ambivalent about this notorious anti-corruption leader. In October 2015 he called for a public protest against corruption allegations involving PM O'Neill. While Anjo said there was 'huge support' for the campaign on social media, few turned up (Ewart, 2015). For many, Anjo has become a figure intimately tied to the politics that they associate with perpetuating corruption throughout the country.

If Anjo or other members of the Coalition were to get into political office, it is likely their critical voices would be silenced. PNG has a history of activists running for political office, and getting into power can significantly disrupt their movements. This occurred after the Sandline Affair when those associated with civil society organisations protesting against the government became politicians themselves. When they became MPs during the 1997 election, previous civil society leaders helped the, allegedly corrupt, late Bill Skate (ABC Four Corners, 2002) become PM and thus lost credibility with their supporters. The once radical NGO, Melanesian Solidarity – who were at the forefront of the Sandline Affair – effectively folded once their leader, the current Governor of NCD, Powes Parkop, was elected in the 2007 election (chapter 4). The lure of political careerism has helped shape political protest in other situations. In 2015, anthropologist Ivo Syndicus suggested that students at the University of Goroka protested against corruption to build their political careers rather than because of their moral opposition to corruption (Syndicus, 2015).

This suggests that the key threat to the Coalition, and groups like it, is not the state's hard power. That is, not its ability to directly confront protestors with force. Rather, it is the state's soft power – expressed through the allure of political office – that shapes the nature of anti-corruption discourse in PNG. This means that more radical anti-corruption movements are likely to continue in PNG, at least for as long as their leadership can withstand the state's political inducements.

Politics and the OC PNG

The OC PNG's independence is ostensibly guaranteed under s217 of the Constitution, which notes that in the performance of its functions 'the Commission is not subject to direction or control by any person or Authority'. For some this ensures the OC PNG's independence while discharging its constitutional duties (Cox, 2004: 91). But political elites continue to shape the nature of the OC PNG and its leadership. One way this is achieved is through the appointment of Commissioners.

The Chief Ombudsman and fellow Ombudsmen are appointed by the Head of State (the Queen of England) who acts in accordance with the advice of an Ombudsman Appointments Committee. As outlined in s217 of the Constitution this committee consists of:

a   the Prime Minister, who shall be Chairman; and
b   the Chief Justice; and
c   the Leader of the Opposition; and
d   the Chairman of the appropriate Permanent Parliamentary Committee, or, if the Chairman is not a member of the Parliament who is recognised by the Parliament as being generally committed to support the Government in the Parliament, the Deputy Chairman of that Committee; and
e   the Chairman of the Public Services Commission.

The inclusion of both the PM and the Leader of the Opposition ostensibly helps ensure impartiality. However, there are signs that this committee is far from free from political influence, as demonstrated by the case of Chief Ombudsman Rigo Lua.

Lua was appointed Chief Ombudsman in mid-2013, at a time when many were questioning the agency's willingness to go after the worst offenders. It had been over a year since the OC PNG had had a Chief Ombudsman, Chronox Manek. Manek was Commissioner from August 2008 to April 2012. He was shot in the arm by unknown assailants in 2009 but continued in the post until, suffering from poor health, passing away in late 2012. On top of this, Ombudsman John Nero was himself referred to the public prosecutor (*PNG Post-Courier*, 2013a) for alleged leadership offences (he passed away before his case was heard). By the time Phoebe Sangetari had been appointed as Acting Chief Ombudsman the OC PNG's capacity had been severely undermined.

Once appointed, Lua quickly turned the organisation around. Soon after his appointment, Lua announced that he and Ombudsman Phoebe Sangetari had issued a directive to freeze progress on a 3-billion-kina (US$1 billion) government loan from Swiss financial services company, UBS. The government used the loan to purchase a 10.01 per cent share in Oil Search – a partner in the country's LNG project. Lua also oversaw the referral of a large number of politicians for the first time in many years. At one point during Lua's reign, the OC PNG had helped in the investigation of around 10 per cent of the 111 MPs, resulting in numerous MPs facing the Leadership Tribunal (ActNow, 2015), an ad hoc tribunal designed to enforce the Leadership Code, which has the power to dismiss, suspend or fine state leaders guilty of misconduct. While many of the issues brought before the Leadership Tribunal were for minor offences – for example, MP Francis Awesa was referred over building a stone wall on his property, while another, Delilah Gore, was found guilty of threatening an airline steward – the perusal of the PM and other MPs within the government showed the OC PNG's renewed determination to investigate powerful politicians. In turn, the political response was swift.

In late 2015 Lua turned 55 years old, giving his political opponents an opportunity to silence him. In PNG, 55 is a significant age for a Commissioner. Section 10 of the Organic Law on the Ombudsman Commission states that a 'person who has attained the age of 55 years shall not be appointed or re-appointed as a member of the Commission and a person shall not be appointed or re-appointed for a period that extends beyond the date on which he will attain the age of 55 years'. This section also allows for reappointments on a case-by-case basis, on the proviso that the Commissioner has not reached the age of 60. Essentially, Rigo's reappointment was possible, but could only occur with the agreement of the Ombudsman Appointments Committee (OAC), which it had provided to previous Chief Ombudsman Ila Geno in 2004.

In Lua's case, however, the OAC simply failed to meet, and has yet to do so at the time of writing. As a result, the opposition leader Don Polye accused

O'Neill of playing politics with the appointment and suggested such decisions contributed to PNG's poor ranking on corruption indices. O'Neill, in turn, accused Polye himself of politicising the appointment process; he stated that the delay was outside of his control, caused by a lack of a quorum. By March 2016 the Chief Ombudsman position had been vacant for ten months. It was becoming increasingly clear that Lua had been sidelined and that the OAC would likely appoint another Chief Ombudsman.

Politicians have also made efforts to curb the powers of the OC PNG through legislative changes. As mentioned, in 2005 TI PNG and the Community Coalition Against Corruption (CCAC) helped to ward off a draft members bill to weaken the OC PNG's powers. Despite this victory, attacks against the institution have continued. In 2010 between 4,000 and 7,000 protesters took to the streets to protest against the 'Maladina Amendments' to section 27(4) of the constitution. The CCAC gathered 20,000 signatures opposing the bill and presented these to the then Leader of the Opposition. Even though the proposed amendment law had passed anonymously through two readings of Parliament, it did not become an Organic Law. While they are increasingly resisted, such attacks on the OC PNG highlight the willingness of politicians to diminish the OC PNG's mandate.

In turn, the threat of political censure looms large over the OC PNG's investigations. This has made the agency, at times, strategic in its approach to investigating politicians. This was evident with Chief Commissioner Geno's parting shot at the then PM, Somare. After an investigation by the OC PNG – which was made public in 2008 – Somare was referred to the Leadership Tribunal for submitting late and incomplete financial statements over 20 years. Geno announced the investigation just before his retirement, leading some to suggest Geno's timing was politically calculated: it meant that he could conduct the investigation without it affecting his career. Somare took little time in attacking the Commissioner for this move. After a mammoth legal struggle Somare was found guilty of 13 charges of misconduct in 2011 (Fox, 2011) but he was given a slap on the wrist: he was merely suspended for two weeks without pay. Geno and previous Commissioners would have known that Somare's financial statements were not right and that he had a constitutional obligation to declare them. But they all turned a blind eye to this omission until a Chief Commissioner, with little to lose, was willing to take on PNG's Grand Chief.

While politics continues to threaten the OC PNG, it has by and large survived the numerous attacks on it. This is in marked contrast to the short-lived, but effective, ITFS. As highlighted in chapter 3, ITFS was established by PM Peter O'Neill in August 2011, to differentiate himself from his predecessor Somare after he controversially replaced Somare as PM. ITFS helped legitimise O'Neill's new government.

The agency quickly secured a number of convictions, including high-profile politicians. The first high-profile case it investigated involved a 10-million-kina (US$3.2 million) deal between infamous businessman Eremas Wartoto and former Minister for National Planning and Monitoring, Paul Tiensten. The payment was made to Wartoto's airline company, Travel Air. Tiensten was found guilty of misappropriation under Section 383A of the Criminal Code and was sentenced to nine years of hard labour, one year short of the maximum sentence. He was also forced to step down from his seat in Parliament. Subsequently, Tiensten was found guilty of defrauding the state by agreeing to a payment to Tol Port Services by his National Planning Department, which resulted in unauthorised transactions. This added three years to his nine-year sentence (*PNG Post-Courier*, 2015). Additional prosecutions of MPs – such as Francis Potape, jailed for 2.5 years for misusing constituency funds and Havila Kavo, sentenced to 3 years for misappropriating funds from a trust account – bolstered the public's support for the ITFS, even though both men's convictions were later overruled by the Supreme Court.

ITFS' initial political support meant that it was able to gain the cooperation of a range of anti-corruption institutions. Indeed, ITFS included senior members of the police force, Public Prosecutor's Office, Internal Revenue Commission, the OC PNG, the Auditor General's Office, the Justice Department, the Treasury and provincial and local-level governments. The Police Department's Fraud and Anti-Corruption Unit, the FIU, along with other agencies which are a part of the National Anti-Corruption Alliance, led the investigations. Inter-agency cooperation was a key reason why ITFS was able to quickly recover tens of millions of kina and helped secure numerous suspensions and arrests of public servants and businesspeople (Fox, 2012).

Despite its early wins, ITFS' tenacious approach led to its downfall. In 2014, it investigated PM O'Neill for his alleged involvement in approving legal fees being paid to a law firm headed by lawyer Paul Paraka. After the arrest warrant for the PM was released, a bitter court battle ensued, which has yet to be resolved. Reprisals against this move were swift. O'Neill declared that the ITFS would be shut down – the fact that the ITFS was not a permanent anti-corruption agency (i.e. it was not inscribed in law) made it easier for him to do so. A court battle over the PM's right to disband ITFS resulted in the agency winning a stay on the decision, effectively giving it a mandate to operate. However, falling out of favour with the PM mortally wounded ITFS. At the end of 2014, Koim revealed that the ITFS had not received any funding from the government, and was close to shutting down (Koim, Walton, & Betteridge, 2014).

Though without funding or political support, at the time of writing the ITFS was still pursuing cases. Sam Koim continued to work on some of the 350 cases of corruption ITFS had registered between 2011 and 2015. Within that time it had secured 12 convictions (Koim, 2015). With the ITFS starved of funds it is unlikely many more will eventuate. The case of ITFS shows that going after politicians in PNG has personal and institutional costs.

Given the threat that politicians have posed to anti-corruption institutions in PNG, some take the view that 'if you can't beat 'em, join 'em'. After retiring from the role of Chief Commissioner, Ila Geno announced he would contest the 2012 elections under the newly formed PNG Constitutional and Democratic Party (PNGCD). The party was Geno's brainchild, and he brought in a number of people engaged in fighting corruption to help run the party – including John Toguata, former Director of Operations in the OC PNG (Hriehwazi, 2011). The party promoted itself as representing 'integrity in leadership', and ran a respectable 39 candidates (in comparison, O'Neill's People's National Congress [PNC] party ran 89 candidates), with Geno running in Rigo Open, a seat in Central Province, against the incumbent Ano Pala, who was re-elected and retained his position as Appointed Minister for Foreign Affairs and Immigration in the then O'Neill–Namah government. The PNGCD won 1 out of 111 seats in Parliament, with Tobias Kulang winning a by-election in August 2011 and then in the general election in 2012 in Kundiawa-Gembogl Open electorate. As the lone representative of this new party Kulang switched to the People's Progress Party and then, in 2015, to the ruling PNC. Clearly, the PNG public does not simply vote for political parties with strong anti-corruption credentials. While this attempt to leverage the OC PNG's reputation for a shot at political office by and large failed, it shows how the position of Commissioner should not be simply read as a bureaucratic post. It is a post that is pregnant with political possibilities.

Discussion and conclusions

This chapter has demonstrated how the real politics of corruption challenged the mainstream view and showed the weaknesses of international approaches to addressing corruption. The Australian aid programme had little room to manoeuvre past its apolitical mandate given PNG's increased leverage over its ex-colonial master. In turn, this results in a cautious and reactionary response to political corruption. By considering corruption a mostly technical issue, as the mainstream approach tends to do, Australia has been poorly equipped to proactively respond to political corruption itself, even though it may be making some inroads in strengthening anti-corruption institutions through garnering political support behind the scenes. This leaves Australia in a position where it is in danger of embodying what David Chandler (2007) labels 'anti-foreign policy', that is: 'attempts to use the international sphere as an arena for self-referential statements of political mission and purpose, decoupled from their subject matter, resulting in ad hoc and arbitrary foreign policy-making' (362). In Australia's case, policy statements are decoupled from the politics that enables corruption, and with Australia's response mostly determined by external actors (such as Sam Koim and Global Witness), it has also resulted in somewhat arbitrary diplomatic and programmatic responses to corruption.

Still, the Australian aid programme has supported actors with greater room to manoeuvre, such as TI PNG and the OC PNG, indicating that the agency plays a role in addressing political corruption beyond the restrictions of its official mandate. TI PNG's leaders have taken risks in responding to corruption, and in so doing have rubbed up against the organisation's non-partisan mandate. Driven by passionate individuals, the organisation has been impactful by occasionally drawing on the critical perspective and questioning the decisions of political leaders, while not directly associating them with corruption per se. In turn, the extent to which the organisation has been able to challenge political corruption has oftentimes had more to do with personality than policy. Ultimately, however, the organisation works within the bounds of its international mandate, and normatively positions the problem of corruption as one of laws, rules and the norms of the state.

This chapter has shown that Papua New Guineans have played a fundamental role in addressing and highlighting political corruption. But international responses have left many wanting for more meaningful political engagement. Organisations like the Australian aid programme and TI PNG, with their tendency to cautiously engage in politics and not 'name names', are not equipped to engage with the political conclusions people draw when discussing corruption. Some have, in turn, supported groups like the Coalition, despite question marks around their connections to prominent politicians. By focusing on the failures of individual politicians the Coalition tapped into the growing frustration that many in the country had (and continue to have) with their political leaders. In doing so, the Coalition filled a gap created by the politically cautious anti-corruption industry. Its political engagement highlights its willingness to strongly reflect the critical perspective when challenging those in political power.

There are aspects of the Coalition's approach to fighting corruption that are harmful. As explained in chapter 4 they have undermined local responses to corruption in PNG, and display a type of xenophobia that does their cause no justice. Still, within the context of PNG's weak civil society, such politicised activists are important because they fill in for development actors who make claims about fighting corruption but are too timid to jump into the political fray. For example, when PM O'Neill dodged an arrest warrant for corruption, and defunded ITFS, Australia – despite its commitment to addressing corruption in PNG and elsewhere – was by and large silent. In comparison, the Coalition's Noel Anjo organised protests, and helped citizens voice their disapproval with the O'Neill government. They are one of the few groups willing to take prominent politicians to task over their corruption, making the Coalition an important (although marginalised) part of PNG's anti-corruption landscape.

The OC PNG is legally mandated to investigate politicians and refer them to the Leadership Tribunal, and its policies and staff reflect the legal and political perspectives of the mainstream view (chapter 4). Yet the effectiveness of the organisation in responding to corruption is strongly tied to the political risks individuals are willing and able to take. This was exemplified by the timing of Ila Geno's prosecution of PM Somare. Strategically managing political attacks can make or break state-based anti-corruption institutions, as shown by the case of ITFS. So the way individuals handle the politics of their positions plays a crucial role in determining the continuity and success of state-based anti-corruption institutions like the OC PNG.

The chapter also showed that national organisations were far more individually invested in politics than their international counterparts. The Coalition were strongly connected to politicians, with one of their leaders running for office and cosying up to the opposition. Similarly, key leadership positions in the OC PNG cannot be read in strictly Weberian terms (i.e. where public office is separated from private interest). Rather, these roles – with their politically popular messages about anti-corruption – have the potential to catapult leaders into political careers, as demonstrated by Ila Geno's foray into political campaigning. These national responses to corruption are not simply motivated by concerns about corruption – they are also tied up in political ambition.

In sum, responses to political corruption in PNG are often personal rather than institutional. Institutional policies and mandates can get in the way of responding to political corruption. Ultimately, political corruption may only be addressed when leaders are supported, and given space, to play politics. Brave souls, rather than Weberian bureaucrats or nervous civil society actors, are in turn essential in ensuring effective responses to political corruption in the country.

Notes

1 The inquiry was tabled in Parliament in 2010 and highlighted widespread corruption in the Department of Finance (Webster, 2010).
2 An objective stressed to me by a number of senior associates of the organisation.

7 Conclusions

Appraising anti-corruption efforts in a weak state

Introduction

The mainstream and alternative views on corruption provide competing viewpoints on how corruption is defined, caused and should be addressed. The former focuses on corruption within the state and the latter looks outside the state to the structural factors that frame corruption, and the power relations embedded within anti-corruption discourse, policy and programmes. Competing narratives about the relevance of these views play out within different societies, and PNG is no exception. Despite the cultural, economic, social and political challenges facing the country, most academics writing about corruption in PNG reflect the mainstream view.

While there are exceptions to this rule, the international anti-corruption industry has come to reflect the mainstream view. This global industry has helped spread responses to corruptionism which reflect this view around the developing world. PNG is just one outpost of this global 'movement', which grew out of a disenfranchisement with the practices of donors in Africa, and helped to give greater purpose to a development industry whose structural adjustment policies had become unpopular and ineffective (Szeftel, 1998). However, the international anti-corruption industry operates in countries where social, economic, cultural and environmental complexities work to challenge the certainties of the mainstream view on a daily basis. States in developing countries are often weak, and the separation between public and private spheres does not exist. In these weak states, non-state actors – communities and the private sector – play an important role in delivering goods, services and social protection systems. This suggests that the mainstream view's approach to understanding and responding to corruption may be inadequate within these contexts.

This book focused on the case of PNG to better understand how the mainstream and alternative views are reflected by presenting findings on local, national and transnational corruption (chapter 3), national anti-corruption organisations (chapter 4), international anti-corruption organisations (chapter 5) and responses to political corruption (chapter 6). This concluding chapter highlights some of the key implications from these findings and what they mean for both theories about corruption and the effectiveness of responses to corruption in PNG. This concluding chapter draws together the main findings of the book by examining four key themes. First, the chapter looks at what the findings mean for the legitimacy of anti-corruption organisations in PNG. The second section examines what the Papua New Guinean Ways should mean for fighting corruption in the country – and the ongoing war on corruption. The third section provides a reflection on how more meaningful responses to corruption might be approached. Finally, implications arising out of the uneven response to powerful elites in PNG are discussed.

Are anti-corruption actors in PNG legitimate?

At its most abstract, 'legitimacy' relates to 'the moralization of authority, the moral grounds for obedience to power, as opposed to grounds of self-interest or coercion' (Parkinson, 2003: 182, citing Crook [1987] and Poggi [1978]). When decisions or regimes are 'legitimate' those who refuse to accept them can be coerced into following them, on the grounds that their refusal is illegitimate (Parkinson, 2003). This definition suggests that its core legitimacy is about institutions, individuals and organisations conforming to expectations (legal or otherwise). The question emanating from this discussion then is: do anti-corruption organisations have the 'moral authority' to call on citizens to reject and meaningfully resist corruption (which is a core part of what they do)?

Those critical of the anti-corruption industry suggest they may not. According to some critical theorists (Haller & Shore, 2005b; Sampson, 2005, 2010; Bukovansky, 2006), international anti-corruption organisations represent a form of Westernisation that jars against local understandings of corruption in developing countries. However, international agencies suggest that their mainstream view about corruption legitimately reflects the concerns of citizens in developing countries. TI, for instance, stresses that its 'movement' gains legitimacy from local chapters around the world (chapter 5). Academics have also supported this view (Noonan, 1984; Klitgaard, 1991; Lambsdorff, 2007).

Supporting those who consider anti-corruption activities in line with local concerns, this book has shown that there are significant similarities between many rural citizens and those associated with international and local anti-corruption organisations. Chapter 3 showed that enfranchised groups were more likely to reflect the mainstream view, while chapters 4 and 5 showed that the overall approach of both national and international anti-corruption agencies also reflects this mainstream view.

That the concerns of international anti-corruption organisations are not foreign to many Papua New Guineans suggests that framings of corruption are not unique to the anti-corruption industry – at some level, many people recognise that transactions that compromise the rules of the public sector, or laws of the land, constitute corruption. What the findings of chapter 3 highlight, however, is that anti-corruption agencies in PNG reflect the concerns of the enfranchised. Marginalised groups were less likely to reflect the mainstream view and were more willing to articulate the potential benefits of corruption (chapter 3). These responses suggest messages of anti-corruption organisations in and of themselves had little moral authority among more marginalised groups because they failed to address the structural constraints framing corruption.

Most anti-corruption activity in PNG is predicated on the assumption that good governance leads to development. Anti-corruption work is focused on educating citizens, overseeing and advising governments, raising awareness, and media campaigns (see chapters 4 and 5). It posits that if citizens say no to corruption and pressure governments to become more accountable and transparent, corruption will lessen, development will eventually materialise and citizens' lives will improve. However, the findings of this book suggest that, in addition to educating Papua New Guineans about the pitfalls of corruption and their obligations to the state, anti-corruption organisations need to prove there is a relationship between development and 'ethical' behaviour. People's marginalisation from 'development', and the weakness of the PNG state, make them more likely to support acts that anti-corruption organisations consider corrupt. This is because many respondents are not convinced that Western standards of right conduct bring development.

The hunger that many in PNG have for a type of development which prefers resource allocation – often over concerns about transparency, accountability and corruption – is evident in the growth of constituency funds in the country. Anti-corruption organisations – particularly TI PNG and its partners (see chapter 5) – have rallied against what they call 'slush funds' for MPs. They are right to be concerned; however, this concern can get in the way of understanding the reason why these funds are popular with both constituents and politicians: it is in part because they provide a means of distributing government money that bypasses weak and dysfunctional government structures. This results in MPs (or political 'big-men') taking direct responsibility for development. Constituency funds have been found to be spent with little transparency, potentially fraudulent, and unfairly distributed (Auditor-General's Office of Papua New Guinea, 2014). Yet, this system of distribution has been supported by large swathes of the population because it mimics the more traditional mechanisms of governance, in ways that approaches advocated by the mainstream view (separation of politics from government, private from public) do not.

If anti-corruption organisations want to reach out to marginalised people (and it is recommended that they should), they will need to convince them that development manifests when 'corruption' is curtailed. The findings of this book suggest that this may occur when anti-corruption organisations are involved in projects that show tangible results. This will mean placing greater emphasis upon providing traditional forms of development: building schools, hospitals, roads, providing teachers, increasing security and so on. Monitoring for bribery, nepotism and other issues will be an important adjunct to these activities. In other words, anti-corruption activity needs to become a means to development, not an end in and of itself. It needs to move away from framing corruption as the enemy in a war, towards more nuanced understandings and responses.

This approach will, no doubt, change the nature of the anti-corruption industry. In PNG, the funding that the Australia aid programme channels into governance (which reached almost 50 per cent of the budget in 2008–09) should be channelled to other sectors that provide a more immediate impact on addressing poverty through provision of social services. In addition, anti-corruption activity must be evaluated in terms of its development outcomes – by showing how much better health facilities/schools operate because of anti-corruption activity. Some of these gains will be difficult to measure, but efforts should be made to do so.

This means that anti-corruption organisations that are predominantly engaged in advocacy, such as TI PNG, should link with development organisations and show that anti-corruption efforts directly lead to tangible outcomes, particularly for rural and marginalised communities. This is a view supported by the Australian aid programme's ODE, whose review of anti-corruption programmes in three countries, including PNG, stated that: 'NGOs that focus exclusively on advocacy often lack the development experience to be credible' (Office of Development Effectiveness, 2007: 12). Organisations such as TI PNG need to prove that reducing what they consider to be corruption directly benefits marginalised communities. Far too much of anti-corruption work in PNG, at the time of fieldwork and of writing, assumes the opposite: that many Papua New Guineans are corrupt and that they need to be taught, pressured and cajoled on how not to be.

Rehabilitating the diminished dreams of PNG's founding leaders

There has been a significant shift in the way elites in PNG have sought to understand and respond to corruption. Some of the nation's founding leaders were keen to highlight the pernicious nature of corruption while chastising the corruption of foreign governments, and Western transnational corporations (chapter 4). The constitution acknowledges that PNG's various cultures should be drawn upon to help bring meaningful development. Academics have been champions of PNG-led solutions to development challenges in years gone by. A year before the country's independence, Hank Nelson (1974: 230) wrote that:

> In the long term it is the people of Niugini who make one confident. They possess a courtesy, imagination and pragmatic strength to provide their own solutions.

However, since independence in 1975, there has been a marked shift away from seeing Papua New Guinean cultural values as a strength to fight corruption;

more recently culture has been framed as a key cause of corruption – as reflected by the mainstream view.

This book showed that this shift was embodied by the different approaches of the Coalition and its predecessor, Melanesian Solidarity (MelSol) (chapter 4). The Coalition's negative nationalism was in stark contrast to MelSol's left-leaning nationalism, which saw PNG culture and the state as more virtuous than foreign states and cultures. While those in the OC PNG were keen to see more culturally sensitive approaches to anti-corruption activities, the organisation itself favoured a more legally oriented approach to addressing it. So over the past 40 years of PNG's independence the notion that Papua New Guineans can solve problems such as corruption has been significantly degraded, because mainstream concerns about corruption have increased at the same time that the state's ability to provide development has deteriorated. In turn, culture and good governance are considered to be like oil and water, they just don't mix. This view is promoted by national and international organisations operating in PNG, who have reinforced an interpretation of corruption which rides roughshod over the structural concerns facing marginalised communities. While it is vital that corruption – in the normative sense of the term – is addressed in PNG, there is a need to rethink the ways anti-corruption efforts can better reflect the Papua New Guinean Ways.

Why?

For a start Papua New Guinean Ways have proven to be more effective than the state in delivering development and providing a social protection system. Communities in PNG still rely on informal social protection systems to fill in for a dysfunctional and weak state. Clans are governed by chieftain (hereditary leadership) or 'big-man' (with authority gained by men deemed most socially, culturally, politically and economically capable) systems, and for many are the primary source of social protection and identity. They provide security and meaning in the face of poor services, unemployment, disease and poverty. And these traditional governance systems have lessons for anti-corruption organisations. While acknowledging the challenges of generalising in such a diverse country, in traditional governance systems resources were mostly produced and consumed openly. People in the village had a good idea of how many pigs and fruit trees others 'owned' and what was growing in others' gardens. Traditional systems emphasised the importance of reciprocity and gift exchange. These systems of governance were therefore far more transparent and egalitarian than the systems of modernity that are attempting to replace them. Indeed, the modern economy places greater emphasis on, and provides more opportunities for, opaque individual accumulation. Citizens are encouraged to think and act as individuals and earn money that is hidden in bank accounts and spent on modern consumer goods which are to be individually consumed. This does not mean that we should become overly romantic towards PNG's past; the benefits of the modern state have been numerous (reducing tribal fights, improving health and education are some of the advantages). However, PNG's diverse cultures and traditions should not be forgotten in the ongoing efforts to address corruption in the country.

In addition, the state has itself been shaped by culture in such a way that it no longer resembles – if it ever did – the Weberian institution so longed for by development and anti-corruption actors. There are signs that the state has become increasingly captured by the cultural values and tribal allegiances of its citizens. Indeed, there are signs that the state has become transformed by these interactions and should be thought of as a 'hybridised' institution as a result. Sinclair Dinnen suggests the PNG state has been subject to a process of 'upward colonisation' through 'social forces emanating from the most local levels [that] have penetrated even the commanding heights of the modern state' (Dinnen, 1997: 10).

This state of affairs is not unique to PNG. Indeed, most countries of the global South constitute hybrid forms of legitimacy. In many countries, people and political elites try to find ways to 'marry' indigenous customary and communal institutions of governance on the one hand and introduced Western state (and civil society) institutions on the other (Clements, 2008). PNG is a prime example of this. The Papua New Guinean traditional, charismatic and rational-legal forms of legitimacy exist side by side.

Connecting formal systems of governance with local realities can help to tap into community resilience and problem-solving capacities (Organisation for Economic Co-operation and Development, 2010). In Bougainville, for instance, Clements (2008, cited in Organisation for Economic Co-operation and Development, 2010: 42) suggests that post-conflict state-formation mixes both traditional and legal-rational forms of legitimacy. He suggests:

> Direct democratic elements stemming from the customary sphere are incorporated into the formal processes of liberal democracy (*e.g.* voter-initiated legislation and plebiscites or the recall of members of parliament), which enhances the legitimacy of these processes. The council of elders, which provide the mainstay of political order, are legal institutions but allow for local variations in the election/selection of members, and include traditional chiefs and elders together with representatives of societal groups (women, youth, and the churches). They thus combine traditional and legal-rational authority.

Similarly, in his book, *The Unseen City*, anthropologist Michael Goddard highlights how PNG's village court system helps bring about justice in line with local expectations through the application of traditional forms of dispute settlement, and modern laws and courts (Goddard, 2005). In other words, the potential for a more nuanced and culturally relevant (alternative) view on corruption within the PNG context stems from the fact that culture is a central part of state institutions already, and has, in some cases, augmented the effectiveness of these institutions. Anti-corruption actors in PNG would do well to take note of the potentials of these more hybrid forms of governance.

Re-establishing the alternative view on corruption in PNG

It is understandable that academics and policy makers are cautious about writing from an alternative view, lest they provide further ammunition to elites to engage in malfeasant activities. So, it is worthwhile noting that much scholarship from the moral and critical perspectives seeks to increase the accountability of elites – not to excuse their actions by relativising corruption. For example, Harrison (2007: 672) argues that challenging mainstream assumptions about the nature of corruption 'is not to present a relativistic position where, for example, corruption is relabelled as "gift giving" and thus excused as culturally acceptable'. 'Who could argue with ... the exposure of corrupt and money-grabbing politician[s] or bureaucrats?' she rhetorically asks (673). Indeed, many critical scholars argue that the existing anti-corruption industry is not focusing enough on exposing the corruption of elites. For example, Hindess (2005) argues that the structural-functionalist tendency of mainstream approaches to addressing corruption placates elites, rather than calls them to account – a finding that resonates with chapter 6. Scholars writing from the alternative view seek to understand, and correct, what they see as the skewed nature of the narratives about corruption – not excuse the corrupt behaviour of elites.

The case for greater alternative scholarship is also strengthened when considering how well anti-corruption efforts, guided by the mainstream view, have fared in PNG. While a few individual anti-corruption campaigns have proved successful (e.g. Pelto, 2007), many doubt the overall effectiveness of anti-corruption organisations' initiatives in the country. For example, the late Mike Manning, past chairman of the local chapter of TI PNG, wrote in its 2007 annual report that: 'Many people ask what effect does TI PNG have on the level of corruption in PNG and the answer is not very clear' (Transparency International Papua New Guinea, 2008a: 4). Papua New Guinean academics Okole and Kavanamur suggest that, given the nature of corruption in PNG, it is difficult to say whether organisations such as the OC PNG 'have made marked differences at all to deter corruption' (2003: 15). As Gebel (2012) suggests in her reflection on TI's global anti-corruption discourse, this does not mean we should dismiss these efforts altogether – but it does show the limitations of these mainstream approaches in PNG (and elsewhere).

Scholarship from the alternative view is not a silver bullet that will address concerns about corruption in the country. But it may help to shift thinking about corruption towards a more nuanced understanding of the issue. The broader development industry itself has a precedent for this. Critiques of the development industry in the 1990s – particularly from the post-development school (Escobar, 1984, 1995; Ferguson, 1994) – helped to shift the discourse and practice of donors and NGOs (Nederveen Pieterse, 2000; Jakimow, 2008). Policy makers and activists can only be aided by having access to and incorporating a broader range of perspectives about corruption in PNG.

There are some in PNG who promote the alternative and mainstream views, suggesting that the answer to PNG's problems with corruption could reside in learning from both. In the literature on corruption in PNG this is evidenced in the writings of Papua New Guinean scholar and at the time of writing Secretary for Department of Higher Education, Research, Science and Technology, David Kavanamur. His discussion paper titled 'The Interplay between Politics and Business in Papua New Guinea' reflects the economic perspective (of the mainstream view) by calling for further privatisation to reduce 'rent seeking'. His article with Henry Okole (2003), however, stresses the importance of local responses to corruption, a reflection of the moral perspective (of the alternative view). The intermingling of these perspectives is even more apparent in public discussions about corruption in PNG where corruption is interpreted in a wide variety of ways. This is evidenced by responses to key political events that have triggered widespread concern about corruption (such as the Sandline Crisis of 1997 – see Dorney, 2001).

While there has been a flourishing of the alternative view in the international literature (Michael, 2004; Brown & Cloke, 2011), writings from the alternative view in PNG are flailing. In PNG, there has been little analysis of the multi-faceted anti-corruption industry (apart from an examination of AusAID – see Temby [2007]), nor has there been much analysis of how discourses of corruption in PNG relate to local understandings. This is surprising given the long history of place-specific anthropological inquiry in the country – with anthropologists contributing to, if not dominating, the alternative view in the international literature.

In PNG the nascent formulations of local alternatives to the mainstream view are encouraging, but it is yet to be seen how these views might be translated into broad-scale anti-corruption institutions and initiatives in a way that is both sensitive to local understandings of corruption, and uniform enough to meaningfully inform national anti-corruption policies. While major donors to PNG, like the Australian aid programme, do encourage citizens to collect and disseminate information about the costs of corruption, there is a large divide between Okole and Kavanamur's suggestion that the *wantok* system should be scaled up, and the overriding concern of donors in the Pacific that mostly reflects the mainstream view (Huffer, 2005; Hutchinson, 2005).

There is much more to discuss about how more nuanced approaches to anti-corruption discourse and activities might take place. A normative exami-nation of the ways in which resources could be allocated and narratives changed is beyond the scope of this book. I will leave it to others to turn the analysis here into more decisive recommendations. However, the findings presented would broadly support national policy makers, activists and academics more critically examining the potential for the alternative view to be incor-porated into anti-corruption activity. Such processes could be supported by transnational actors who are willing to provide critical international support to these local and national efforts.

Power, politics and anti-corruption

International and local anti-corruption organisations' work is ostensibly about shifting relations of power and bringing about a fairer world. According to this view, reducing the ability for the powerful to take away state resources through corruption will help bring about prosperity and development. Through keeping the powerful in check, a new world is possible: one where corruption does not eat into funds that would provide health care, education and other markers of development. However, in PNG the powerful are mostly free to do as they please. They have, all too often, been able to dodge the purview of anti-corruption actors, calling into question the effectiveness of anti-corruption efforts.

Chapter 6 showed that the Australian aid programme struggled to meaningfully respond to political corruption, particularly in addressing the transnational transfer of funds from Australia to PNG. TI PNG and the OC PNG were at times cautious, even though they sometimes critiqued political corruption. It was the Coalition who provided a more direct and confrontational response to political corruption in PNG, particularly through their critique of Prime Minister Somare. As AusAID, TI PNG, and the OC PNG were cautious in their approach to addressing political corruption, the Coalition's response was important as they were able to speak truth to power like no other anti-corruption organisation in the country.

Without a strong coalition of anti-corruption actors who are willing to take on powerful politicians and other elites over corruption, the powerful are likely to be largely untroubled by national and international efforts. During the past few years in PNG such a coalition has failed to materialise. The Coalition were the only group willing to make sustained attacks on PNG's politicians over accusations of corruption. Given the precarious and political nature of the Coalition it is unlikely such critique will last, unless there is a broader and longer-term movement against political corruption which is also not afraid to 'name names'.

Another set of powerful actors, those in the private sector, have come to play an important part in facilitating corruption in PNG (chapter 3). Yet not enough has been done to address corruption in this sector, with most anti-corruption efforts focused on government corruption. The anti-corruption industry needs to do much more to address and understand private sector corruption. Chapter 3 showed that the private sector has long been associated with the rise of corruption in PNG. The Coalition was most active in addressing private sector corruption. Their concern about Asian businesses led to an anti-Asian protest-cum-riot in 2009 (chapter 6). This violent confrontation is an unconstructive way of dealing with the problem, leading, as it did, to death and destruction. More constructive ways of addressing private sector corruption need to be sought by anti-corruption organisations in PNG. While all organisations acknowledge the role the private sector plays in corruption, there is a lack of effort to monitor and raise awareness about corruption involving transnational organisations. This is particularly the case with Western corporations, who were often considered as anti-corruption champions – in line with global discourse (Hindess, 2005) – by both national and international organisations. Overlooking the role of the Western private sector in corruption fails to acknowledge the other half of corrupt transactions. It also fails to appreciate the role these organisations play in delivering development and acting like states – particularly in areas where the PNG state is weak. It is in these areas where logging and mining companies can provide traditional state functions – policing, education and health facilities – and are thus more likely to be involved in corruption.

The transnational nature of corruption involving PNG politicians, middlemen and others (chapter 3) also highlights deficiencies in mainstream approaches to addressing corruption. Focusing on corruption that occurs within the nation-state – as the mainstream view tends to – can overlook the transnational relationships that help facilitate corruption. In turn, Australia struggles to adequately respond to accusations that it is a harbinger for the proceeds from political corruption. Insights from the alternative view's critical perspective are important here. In particular, neo-Marxist scholars (Hindess, 2005; Brown & Cloke, 2006, 2011; Rocha et al., 2011; Murphy, 2011) have urged for analysis of anti-corruptionism that accounts for the processes of capitalist commodity production and trade. Their call for a form of analysis which is more sensitive to the processes of economic globalisation should be welcomed in PNG.

Thinking and acting politically in regards to corruption requires a shift in thinking in PNG and across the development industry. Alina Mungiu-Pippidi notes that donors across the world shy away from tackling 'political anti-corruption' (2013: 114). As this book has found, she argues that donors prefer to train government officials. Such technocratic approaches need to be seriously rethought. Given the transnational linkages between the international development industry, it is critical that thinking politically about corruption not be contained to PNG, but be a broader and more global discussion involving critical scholars, activists and others.

In sum, this book provides a more nuanced picture of the potentials and pitfalls of anti-corruption by examining how alternative and mainstream views on corruption were reflected by anti-corruption organisations and Papua New Guinean citizens. A key challenge for academics and policy makers in the coming years will be how to bridge the gap between Western and local understandings of and solutions to corruption. This will mean redirecting resources and reframing engagements. The process should also include giving local scholars and citizens space to develop their ideas about how corruption is understood and can be meaningfully addressed. Allowing for such space, time and resources may be difficult for anti-corruption policy makers, who are used to responding to corruption from a mainstream view and measuring results accordingly. But without doing so, responses to corruption will continue to miss their mark, and result in frustration among policy makers and citizens who daily read and hear about corruption but see little tangible evidence that it is being effectively combatted.

The book highlights how a geographical engagement on corruption that is sensitive to place, space and, in particular, scale can provide important insights on this issue. During writing this book there have been signs that geographers are becoming increasingly interested in critically examining the issue of corruption. In 2014, a special session on corruption was a part of the annual meeting of the Association of American Geographers – organised by Sapana Doshi and Malini Ranganathan. The session evoked much interest. This interest should be encouraged as it is important that geographers and others concerned with space, scale and place are engaged in the unravelling of this contested concept. Doing so should move academics and practitioners closer to understanding and addressing popular concerns about corruption and its impacts.

References

AAP. (2007). Downer Blasts Corrupt Pacific Leaders. 11 May. *Sydney Morning Herald*. Retrieved 8 August, 2014, from http://www.smh.com.au/news/National/ Downer-blasts-corrupt-Pacific-leaders/2007/08/08/1186530439769.html

Abal, S. T., & Smith, S. (2010). Joint Statement. Paper presented at the Papua New Guinea-Australia Bilateral Meeting, Alotau International Hotel, Alotau. Retrieved 24 December, 2013, from http://www.foreignminister.gov.au/releases/2010/fa- s100709a.html

ABC Four Corners. (2002). The Insider: A Banker's Tale of Big Money, Political Manipulation and Failure. Retrieved 24 November, 2016, from http://www.abc.net. au/4corners/stories/s590512.htm

ABC News. (2015). Papua New Guinea PM Says Aid Money Wasted on 'Middlemen'; Calls for Development Support Rethink. 4 August. Retrieved 2 September, 2015, from http://www.abc.net.au/news/2015-08-03/papua-new-guinea-plans-rethink-of-development-support-delivery/6667642

ABC News. (2016). PNG Removes Foreign Advisers with 15 Australian Government Aid Positions Targeted. 6 January. Retrieved 16 January, 2016, from http://www.abc. net.au/news/2016-01-06/png-removes-foreign-advisers/7070344

ABC Radio Australia. (2011). PNG Public Accounts Committee Criticises Accounting System. Retrieved 17 June, 2011, from http://www.radioaustralia.net.au/pacbeat/ stories/201105/s3229166.htm

ACIL Tasman. (2009). *PNG LNG Economic Impact Study: An Assessment of the Direct and Indirect Impacts of the Proposed PNG LNG Project on the Economy of Papua New Guinea*. Melbourne: ACIL Tasman.

ActNow. (2015). Updated: The MP's Charged, Waiting Sentence or Imprisoned. 13 October. *ActNow*. Retrieved 20 April, 2017, from http://actnowpng.org/blog/blog-entry-updated-mps-charged-waiting-sentence-or-imprisoned

Aid Watch. (2005). *Boomerang Aid: Not Good Enough Minister!: Response to Australian Foreign Minister Downer's Comments on Boomerang Aid*. Sydney, Australia: Aid Watch.

Ali, A., & Isse, H. (2003). Determinants of Economic Corruption: A Cross-Country Comparison. *Cato Journal*, 22(3), 449–466.

Allen, M., & Hasnain, Z. (2010). Power, Pork and Patronage: Decentralisation and the Politicisation of the Development Budget in Papua New Guinea. *Commonwealth Journal of Local Governance*, 6, 7–31.

Andvig, J. C., & Fjeldstad, O.-H. (2001). *Corruption: A Review of Contemporary Research*. Bergen, Norway: Chr. Michelsen Institute, Development Studies and Human Rights.

Anechiarico, F., & Jacobs, J. B. (1996). *The Pursuit of Absolute Integrity: How Corruption Control Makes Government Ineffective*. Chicago, IL: University of Chicago Press.

Arriola, L. (2009). Patronage and Political Stability in Africa. *Comparative Political Studies*, 42(10), 1339–1362.

Asian Development Bank. (2004). *Anti-Corruption Policies in Asia and the Pacific: The Legal and Institutional Frameworks for Fighting Corruption in Twenty-One Asian and Pacific Countries*. Manila, Philippines: Asian Development Bank.

Asian Development Bank. (2015). *Pacific Economic Monitor*. July. Mandaluyong City: Asian Development Bank.

Asian Development Bank. (2016). *Asian Development Outlook 2016 Update: Meeting the Low-Carbon Growth Challenge*. Manila: Asian Development Bank.

Auditor-General's Office of Papua New Guinea. (2014). *District Services Improvement Program*. Port Moresby: Auditor-General's Office of Papua New Guinea.

AusAID. (2006a). *AusAID Annual Report*. Canberra: AusAID.

AusAID. (2006b). *Democratic Governance in Papua New Guinea: Draft Strategy and Program Concept Design for Peer Review*. Canberra: AusAID.

AusAID. (2007a). *Annual Report: 2006–2007*. Canberra: AusAID.

AusAID. (2007b). *Scope of Initiative: Anti-Corruption for Development*. Canberra: AusAID.

AusAID. (2007c). *Tackling Corruption for Growth and Development: A Policy for Australian Development Assistance on Anti-Corruption*. Canberra: AusAID.

AusAID. (2008). *Governance Annual Thematic Performance Report 2007–08*. Canberra: AusAID.

AusAID. (2009a). *Australian Agency for International Development Annual Report 08–09*. Canberra: AusAID.

AusAID. (2009b). *Papua New Guinea: Strongim Pipol Strongim Nesen (Empower People: Strengthen the Nation) – Program Design Document*. Canberra: AusAID.

AusAID. (2010). *Annual Thematic Performance Report: Law and Justice 2008–09*. Canberra: AusAID.

AusAID. (n.d.). Brief History of AusAID. Retrieved 4 August, 2010, from http://weba rchive.nla.gov.au/gov/20100724154040/http://www.ausaid.gov.au/about/history.cfm

AUSTRAC. (2008). *Inquiry into the Economic and Security Challenges Facing Papua New Guinea and the Island States of the Southwest Pacific – Submission*. October. Canberra: Australian Transaction Reports and Analysis Centre.

Australian Conservation Foundation & CELCoR. (2006). *Bulldozing Progress: Human Rights Abuses and Corruption in Papua New Guinea's Large Scale Logging Industry*. Port Moresby and Melbourne: Australian Conservation Foundation and CELCoR

Ayius, A., & May, R. J. (Eds.). (2008). *Corruption in Papua New Guinea: Towards an Understanding of the Issues*. Special Publication No. 47. Port Moresby: National Research Institute.

Barker, P., & Awili, S. (2007). The Business and Investment Conditions in PNG in 2007: A Private Sector Survey. 11 December. Paper presented at the Seminar on Private Sector and Investment, Port Moresby.

Barnett, T. E. (1989). *Report of the Commission of Inquiry into Aspects of the Forest Industry*. Port Moresby: Government of Papua New Guinea.

Barth, F. (1978). *Scale and Social Organisation*. Oslo: Universitetsforlage.

Bayley, D. (1966). Effects of Corruption in a Developing Nation. *The Western Political Quarterly*, 19(4), 719–732.

Beder, S. (2001). Privatisation in PNG. *Engineers Australia*, p. 30.

Bishop, J. (2014). The New Aid Paradigm – National Press Club Q and A. 18 June. Retrieved 23 November, 2016, from http://foreignminister.gov.au/transcripts/Pages/2014/jb_tr_140620.aspx

Bohane, B. (2006). Illegal Torres Strait Trade Investigated. Retrieved 9 December, 2016, from http://www.abc.net.au/lateline/content/2006/s1618614.htm

Bratsis, P. (2003). *Corrupt Compared to What?: Greece, Capitalist Interests, and the Specular Purity of the State*. London: The Hellenic Observatory, The European Institute, and the London School of Economics and Political Science.

Brennan, E. (2015). PNG: The World's Biggest Grower in 2015? 14 January. Retrieved 6 October, 2016, from http://thediplomat.com/2015/01/png-the-worlds-biggest-grower-in-2015/

Broadhurst, R. G., Lauchs, M. A., & Lohrisch, S. (2012). Transnational and Organized Crime in Oceania. *SSRN Electronic Journal*, August, 1–19.

Brown, E., & Cloke, J. (2004). Neoliberal Reform, Governance and Corruption in the South: Assessing the International Anti-Corruption Crusade. *Antipode*, 36(2), 272–294.

Brown, E., & Cloke, J. (2005). Neoliberal Reform, Governance and Corruption in Central America: Exploring the Nicaraguan Case. *Political Geography*, 24(5), 601–630.

Brown, E., & Cloke, J. (2006). The Critical Business of Corruption. *Critical Perspectives on International Business*, 2(4), 275–298.

Brown, E., & Cloke, J. (2011). Critical Perspectives on Corruption: An Overview. *Critical Perspectives on International Business*, 7(2), 116–124.

Bukovansky, M. (2006). The Hollowness of Anti-Corruption Discourse. *Review of International Political Economy*, 13(2), 181–209.

Callick, R. (2009). Looters Shot Dead Amid Chaos of Papua New Guinea's Anti-Chinese Riots. 23 May. *The Australian*. Retrieved 2 October, 2010, from http://www.theaustralian.com.au/news/looters-shot-dead-amid-chaos-of-papua-new-guineas-anti-chinese-riots/story-e6frg6no-1225715006615

Callick, R. (2012). PNG Takes in Jakarta Fugitive Joko Tjandra. 21 June. *The Australian*. Retrieved 21 November, 2015, from http://www.theaustralian.com.au/news/world/ png-takes-in-jakarta-fugitive-joko-tjandra/story-e6frg6so-1226403440547

Callinan, R. (2014). Sydney Money Managers Lose $10 Million of PNG Taxpayer Funds. 1 December. *Sydney Morning Herald*. Retrieved 20 November, 2015, from http://www.smh.com.au/business/banking-and-finance/sydney-money-managers-lose- 10-million-of-png-taxpayer-funds-20141126-11upfm.html

Cartier-Bresson, J. (2000). Causes and Consequences of Corruption: Economic Analyses and Lessons Learnt. In Organisation for Economic Co-operation and Development (Ed.), *No Longer Business as Usual: Fighting Bribery and Corruption* (pp. 11–28). Paris: Organisation for Economic Co-operation and Development.

Chan, J. (2016). *Playing the Game: Life and Politics in Papua New Guinea*. St Lucia: University of Queensland Press.

Chandler, D. (2007). The Security–Development Nexus and the Rise of 'Anti-Foreign Policy'. *Journal of International Relations and Development*, 10(4), 362–386.

Chang, H.-J. (2008). *Bad Samaritans – the Guilty Secrets of Rich Nations & the Threat to Global Prosperity*. London: Random House.

Chin, J. (2008). Contemporary Chinese Community in Papua–New Guinea: Old Money Versus New Migrants. *Chinese Southern Diaspora Studies*, 2, 117–126.

Chochrane, L. (2014). Oxfam Accuses Big Four Banks of Funding Companies Accused of Land Grabs. 28 April. *The World Today*. Retrieved 14 November, 2015, from http://www.abc.net.au/worldtoday/content/2014/s3993212.htm

Cirillo, S. (2006). *Australia's Governance Aid: Evaluating Evolving Norms and Objectives*. Asia Pacific School of Economics and Government: Discussion Papers. 5 May. Canberra: Policy and Governance, Australian National University.

Clements, K. (2008). *Traditional, Charismatic and Grounded Legitimacy*. Division State and Democracy Governance Cluster Sector Project: Good Governance and Democracy. Bonn: Deutsche Gesellschaft für Zechnische Zusammenarbeit.

Commission of Inquiry into the Special Agriculture & Business Leases. (n.a.). Summary of Leases Reviewed by the Commission of Inquiry Retrieved 21 November, 2015, from http://www.coi.gov.pg/documents/COI%20SABL/Summary%20-%20 COI%20SABL.pdf

Committee for Development Policy. (2006). *Overcoming Economic Vulnerability and Creating Employment: Report of the Committee for Development Policy for the Eighth Session*. 20–24 March. New York: United Nations.

Commonwealth of Australia. (2008). Port Moresby Declaration. Retrieved 20 April, 2017, from http://www.aph.gov.au/parliamentary_business/committees/house_of_rep resentatives_committees?url=haa/./pacifichealth/report/appendix%20a.pdf

Constitutional Planning Committee. (1974). *Final Report of the Constitutional Planning Committee: Part 1*. Port Moresby: Constitutional Planning Committee.

Cox, O. (2004). *Papua New Guinea's Law and Legal System*. Port Moresby: Pacific Adventist University.

Crocombe, R. (2001). *The South Pacific* (3rd ed.). Suva: University of the South Pacific.

Cuthbert, N. (2008). Call on All to Fight Corruption. 2 June. *PNG Post-Courier*, p. 8.

Davani, C., Sheehan, M., & Manoa, D. (2009). *The Commission of Inquiry Generally into the Department of Finance: Final Report*. Port Moresby: Government of Papua New Guinea.

de Jonge, A. (1998). Ombudsmen and Leadership Codes in Papua New Guinea and Fiji: Keeping Government Accountable in a Rapidly Changing World. Paper presented at the Asia Pacific Constitutions, The University of Melbourne, Melbourne.

de las Casas, G. (2008). Is Nationalism Good for You? *Foreign Policy*, 8 October, 51–56.

de Maria, W. (2005). The New War on African "Corruption": Just Another Neo-Colonial Adventure? Paper presented at the 4th International Critical Management Studies Conference, Cambridge University, Cambridge.

de Maria, W. (2008). Cross Cultural Trespass: Assessing African Anti-Corruption Capacity. *International Journal of Cross Cultural Management*, 8(3), 317–341.

de Sardan, J. P. O. (1999). A Moral Economy of Corruption in Africa? *The Journal of Modern African Studies*, 37(1), 25–52.

de Sardan, J. P. O. (2005). *Anthropology and Development: Understanding Contemporary Social Change*. London: Zed Books.

Department of Foreign Affairs and Trade. (1997). *Part A: Portfolio and Corporate Overview 1996–1997 Annual Report*. Canberra: DFAT.

Department of Foreign Affairs and Trade. (2014a). The 2014–2015 Development Assistance Budget: A Summary. Retrieved 10 July, 2015, from http://dfat.gov.au/about-us/corporate/portfolio-budget-statements/Documents/2014-15-development-assistance-budget-a-summary.docx

Department of Foreign Affairs and Trade. (2014b). *Australian Aid: Promoting Prosperity, Reducing Poverty, Enhancing Stability*. June. Canberra: Commonwealth of Australia, DFAT.

Department of Foreign Affairs and Trade. (2015). Aid Investment Plan Papua New Guinea: 2015–2016 to 2017–2018. 30 September. Retrieved 22 September, 2016, from http://dfat.gov.au/about-us/publications/Pages/aid-investment-plan-aip-papua-new-guinea-2015-16-to-2017-18.aspx

Development Co-operation Directorate. (n.d.). DAC Glossary of Key Terms and Concepts: Technical Cooperation. Retrieved 9 April, 2017, from http://www.oecd.org/dac/dac-glossary.htm#TC

Di Puppo, L. (2010). Anti-Corruption Interventions in Georgia. *Global Crime*, 11(2), 220–236.

Dinnen, S. (1997). *Law, Order and the State in Papua New Guinea*. State, Society and Governance in Melanesia. Canberra: Australian National University.

Dinnen, S. (2001). *Law and Order in a Weak State: Crime and Politics in Papua New Guinea*. Honolulu: University of Hawaii Press.

Dinnen, S., McLeod, A., & Peake, G. (2006). Police-Building in Weak States: Australian Approaches in Papua New Guinea and Solomon Islands. *Civil Wars*, 8(2), 87–108.

Dix, S., & Pok, E. (2009). Combating Corruption in Traditional Societies. In R. I. Rotberg (Ed.), *Corruption, Global Security and World Order* (pp. 239–259). Baltimore, MD: Brookings Institution Press.

Dixon, G., Gene, M., & Walter, N. (2008). *Joint Review of the Enhanced Cooperation Program (ECP)*. March. Governments of Papua New Guinea and Australia.

Dixon, N. (1997a). Melsol MPs Explain Why They Joined PNG Government. Retrieved 27 August, 2012, from http://www.greenleft.org.au/node/14178

Dixon, N. (1997b). PNG: 'The People Stepped In'. 9 April. Retrieved 31 August, 2015, from https://www.greenleft.org.au/node/15108

Dixon, N. (1998). Activist Slams 'Recolonisation' of Pacific. 29 April. Retrieved 20 August, 2012, from https://www.greenleft.org.au/node/17871

Dorney, S. (2001). *Papua New Guinea: People, Politics and History since 1975*. Sydney: ABC Books.

Dorney, S. (2016). *The Embarrassed Colonialist*. Sydney: Penguin.

Downer, A. (1997). *Better Aid for a Better Future: Seventh Annual Report to Parliament on Australia's Development Cooperation Program and the Government's Response to the Committee of Review of Australia's Overseas Aid Program*. Canberra: AusAID.

Dwyer, P. D., & Minnegal, M. (1998). Waiting for Company: Ethos and Environment among Kubo of Papua New Guinea. *The Journal of the Royal Anthropological Institute*, 4(1), 23–42.

Eroro, S. (2008). Parkop Flays Claim of NGO Campaign. 26 June. *PNG Post-Courier*, p. 9.

Escobar, A. (1984). Discourse and Power in Development: Michel Foucault and the Relevance of His Work to the Third World. *Alternatives*, 10(3), 377–400.

Escobar, A. (1992). Imagining a Post-Development Era? Critical Thought, Development and Social Movements. *Third World and Post-Colonial Issues*, 31/32, 20–56.

Escobar, A. (1995). *Encountering Development: The Making and Unmaking of the Third World*. Princeton, NJ: Princeton University Press.

Escobar, A. (2003). Development. In W. Sachs (Ed.), *The Development Dictionary: A Guide to Knowledge as Power* (pp. 6–25). Johannesburg: Witwatersrand University Press, Zed Books.

Ewart, R. (2015). Peaceful Protest against PNG Prime Minister Peter O'Neill Planned for Next Week. 20 October. *ABC News*. Retrieved 23 November, 2016, from http://

www.abc.net.au/news/2015-10-20/peaceful-protest-against-png-prime-minister-peter/6868498

Fallon, J., Sugden, C., & Pieper, L. (2003). *The Contribution of Australian Aid to Papua New Guinea's Development 1975–2000: Provisional Conclusions from a Rapid Assessment*. Canberra: AusAID.

FATF and APG. (2015). *Anti-Money Laundering and Counter-Terrorist Financing Measures: Australia. Fourth Round Mutual Evaluation Report*. Paris and Sydney: FATF and APG.

Fauziah, I. (2009). Chinese Migrants Face PNG Wrath. *101 East*. Retrieved 2 November, 2010, from http://www.youtube.com/watch?v=ZyD0LKOm6dw&feature=channel

Feeny, S. (2005). The Impact of Foreign Aid on Economic Growth in Papua New Guinea. *Journal of Development Studies*, 41(6), 1092–1117.

Ferguson, J. (1994). *The Anti-Politics Machine: "Development", Depoliticization and Bureaucratic Power in Lesotho*. Minneapolis: University of Minnesota Press.

Flanagan, P. (2015). Expenditure in PNG's 2016 Budget – a Detailed Analysis. *Devpolicy Blog*, 2 December. Retrieved 3 August, 2016, from http://devpolicy.org/expenditure-in-pngs-2016-budget-a-detailed-analysis-20151202/

Foster, R. J. (2002). *Materialising the Nation: Commodities, Consumption, and Media in Papua New Guinea*. Bloomington: Indiana University Press.

Fox, L. (2011). Chief Suspended. 27 March. Retrieved 24 November, 2016, from http://www.abc.net.au/correspondents/content/2011/s3174450.htm

Fox, L. (2012). Corrupt 'Mobocracy' Undermining PNG Democracy. 11 May. *ABC News*. Retrieved 7 May, 2015, from http://www.abc.net.au/news/2012-05-11/png-corruption-widespread/4004694

Garrett, J. (2014). PNG's Land Scandal Inquiry Names an Australian-Led Company. 12 February. *ABC News*. Retrieved 22 October, 2015, from http://www.abc.net.au/news/2014-02-12/an-png-land-scandal-inquiry-names-an-australian-led-company/5255528

Gebel, A. (2012). Human Nature and Morality in the Anti-Corruption Discourse of Transparency International. *Public Administration and Development*, 32(1), 109–128.

Giddens, A. (1984). *The Constitution of Society: Outline of the Theory of Structuration*. Cambridge: Polity Press.

Global Witness. (2015). Papua New Guinea Government Hosts Talks on Forest Protection While Turning a Blind Eye to Multi-Million Dollar Illegal Logging in Its Own Backyard. 26 October. Retrieved 9 December, 2016, from https://www.globalwitness.org/en/press-releases/papua-new-guinea-government-turns-blind-eye-to-illegal-logging-its-own-backyard/

Goddard, M. (2005). *The Unseen City: Anthropological Perspectives on Port Moresby, Papua New Guinea*. Canberra: Pandanus Books.

Goel, R., & Nelson, M. (2005). Economic Freedom versus Political Freedom: Cross-Country Influences on Corruption. *Australian Economic Papers*, 44(2), 121–133.

Goetz, A. M. (2007). Political Cleaners: Women as the New Anti-Corruption Force? *Development and Change*, 38(1), 87–105.

Gordon, R. J., & Meggitt, M. J. (1985). *Law and Order in the New Guinea Highlands: Encounters with Enga*. Hanover, NH: University Press of New England.

Gosarevski, S., Hughes, H., & Windybank, S. (2004a). Is Papua New Guinea Viable with Customary Land Ownership? *Pacific Economic Bulletin*, 19(3), 133–148.

Gosarevski, S., Hughes, H., & Windybank, S. (2004b). Is Papua New Guinea Viable? *Pacific Economic Bulletin*, 19(1), 134–148.

Greenpeace. (2005). *The Untouchables: Rimbunan Hijau's World of Forest Crime and Political Patronage*. Amsterdam: Greenpeace.

Gridneff, I. (2010). PNG PM Denies Kidnapping Allegations. 25 November. *Sydney Morning Herald*. Retrieved 24 November, 2016, from http://www.smh.com.au/brea king-news-world/png-pm-denies-kidnapping-allegations-20101125-188jo.html

Griffin, J. (1997). The Papua New Guinea National Elections 1997. *The Journal of Pacific History*, 32(3), 71–78.

Gupta, A. (1995). Blurred Boundaries: The Discourse of Corruption, the Culture of Politics and the Imagined State. *American Ethnologist*, 22(2), 375–402.

Gupta, A. (2012). *Red Tape: Bureaucracy, Structural Violence, and Poverty in India*. Durham, NC: Duke University Press.

Gupta, S., Davoodi, H., & Alonso-Terme, R. (2002). Does Corruption Affect Income Inequality and Poverty? *Economics of Governance*, 3(1), 23–45.

Haley, N., & Zubrinich, K. (2012). *2012 Papua New Guinea General Elections – Domestic Observation Report (Draft) – Prepared for Electoral Support Program (ESP 3) with Funding from AusAid*. 12 October, 2012. Quoted in Department of Foreign Affairs and Trade. (2013). *Australian Aid for Electoral Assistance in Papua New Guinea, Independent Evaluation*. Canberra: Department of Foreign Affairs and Trade.

Haller, D., & Shore, C. (2005a). *Corruption: Anthropological Perspectives*. London: Pluto.

Haller, D., & Shore, C. (2005b). Introduction – Sharp Practice: Anthropology and the Study of Corruption. In D. Haller & C. Shore (Eds.), *Corruption: Anthropological Perspectives* (pp. 1–28). London: Pluto.

Hamilton-Hart, N. (2001). Anti-Corruption Strategies in Indonesia. *Bulletin of Indonesian Economic Studies*, 37(1), 65–82.

Hanlon, J. (2004). Do Donors Promote Corruption?: The Case of Mozambique. *Third World Quarterly*, 25(4), 747–763.

Hanna, R., Bishop, S., Nadel, S., Scheffler, G., & Durlacher, K. (2011). *The Effectiveness of Anti-Corruption Policy: What Has Worked, What Hasn't, and What We Don't Know–a Systematic Review Protocol*. London: EPPI-Centre, Social Science Research Unit, Institute of Education, University of London.

Hardoon, D., & Heinrich, F. (2011). *Bribe Payers Index 2011*. Berlin: Transparency International.

Hardoon, D., & Heinrich, F. (2013). *Global Corruption Barometer 2013: Report*. Berlin: Transparency International.

Harrison, E. (2006). Unpacking the Anti-corruption Agenda: Dilemmas for Anthropologists. *Oxford Development Studies*, 34(1), 15–29.

Harrison, E. (2007). Corruption. *Development in Practice*, 17(4–5), 672–678.

High Commission of the Independent State of Papua New Guinea. (2001). Wages. In *Labour and Human Resources*. Retrieved 10 September, 2009, from http://www.pngcanberra.org/trade/lhr.htm

Higuchi, Y. (2015). The President's Message. Retrieved 27 July, 2015, from http://www.kumagaigumi.co.jp/english/corp/message.html

Hill, L. (2012). Ideas of Corruption in the Eighteenth Century: The Competing Conceptions of Adam Ferguson and Adam Smith. In M. Barcham, B. Hindess, & P. Larmour (Eds.), *Corruption: Expanding the Focus* (pp. 97–112). Canberra: Australian National University E Press.

Hindess, B. (2005). Review Essay: Investigating International Anti-Corruption. *Third World Quarterly*, 26(8), 1389–1398.

Holmes, L. (2006a). Networks and Linkages: Corruption, Organised Crime, Corporate Crime and Terrorism. Paper presented at Workshop 4.2: Informal Practice and Corruption: How to Weaken the Link, International Anti-Corruption Conference 12, Guatemala City.

Holmes, L. (2006b). *Rotten States? Corruption, Post-Communism and Neoliberalism.* Durham, NC: Duke University Press.

Hope, K. R., & Chikulo, B. C. (2000). Introduction. In K. R. Hope & B. C. Chikulo (Eds.), *Corruption and Development in Africa: Lessons from Country Case-Studies* (pp. 1–13). New York: St Martin's Press.

Howes, S. (2016). Scaled Down. The Last of the Aid Cuts? *Devpolicy Blog*. Retrieved 4 May, 2016, from http://devpolicy.org/scaled-last-aid-cuts-20160504/

Howes, S., Mako, A., Swan, A., Walton, G., Webster, T., & Wiltshire, C. (2014). *A Lost Decade? Service Delivery and Reforms in Papua New Guinea 2002–2012.* Canberra: The National Research Institute and the Development Policy Centre, Australian National University.

Howes, S., & Negin, J. (2014). The New Aid Paradigm: Is It New, and What Does It Do for Aid Reform? *Devpolicy Blog*, 19 June. Retrieved 10 December, 2015, from http://devpolicy.org/the-new-aid-paradigm-is-it-new-and-what-does-it-do-for-aid-reform-20140619/

Hriehwazi, Y. (2011). Geno and Kuri Form Party. 6 January. *PNG Post-Courier*, p. 6.

Huffer, E. (2005). Governance, Corruption and Ethics in the Pacific. *The Contemporary Pacific*, 17(1), 118–140.

Hughes, H. (2003). Aid Has Failed the Pacific. Issue Analysis No 33. 7 May. Retrieved 26 June, 2007, from http://exkiap.net/articles/cis20030507_failed_aid/ia33.htm

Hughes, H. (2004). *Can Papua New Guinea Come Back from the Brink?* Sydney: Centre for Independent Studies.

Huntington, S. P. (1968). *Political Order in Changing Societies.* New Haven, CT: Yale University Press.

Huntington, S. P. (1989). Modernization and Corruption. In A. J. Heidenheimer, M. Johnston, & V. T. LeVine (Eds.), *Political Corruption: A Handbook* (pp. 377–388). New Brunswick, NJ: Transaction Publishers.

Hutchinson, F. (2005). *A Review to Donor Agency Approaches to Anti-Corruption.* Asian Pacific School of Economics and Government Discussion Papers, February. Canberra: Asian Pacific School of Economics and Government, Australian National University.

Independent Commission Against Corruption. (1994). *Unravelling Corruption: A Public Sector Perspective.* Sydney: ICAC.

Independent Review Team. (2010). *Review of the PNG-Australia Development Cooperation Treaty (1999).* 19 April. Canberra and Port Moresby: AusAID.

Independent State of Papua New Guinea. (1975). *Constitution of the Independent State of Papua New Guinea*, 1–140.

Interpol. (2009). Tjandra, Joko Soegiarto. Retrieved 26 November, 2015, from http://www.interpol.int/notice/search/wanted/2009-21489

Ives, D., Midire, C., & Heijkoop, P. (2004). *Governance in PNG: A Cluster Evaluation of Three Public Sector Reform Activities.* Evaluation and Review Series.Canberra: AusAID.

Jackson, M., & Smith, R. (1996). Inside Moves and Outside Views: An Australian Case Study of Elite and Public Perceptions of Political Corruption. *Governance*, 9(1), 23–42.

Jain, A. K. (2001). Corruption: A Review. *The Journal of Economic Surveys*, 15(1), 71–121.

Jakimow, T. (2008). Answering the Critics: The Potential and Limitations of the Knowledge Agenda as a Practical Response to Post-Development Critiques. *Progress in Development Studies*, 8(4), 311–323.

Jayasuriya, K. (2002). Governance, Post-Washington Consensus and the New Anti-Politics. In T. Lindsey & H. Dick (Eds.), *Corruption in Asia: Rethinking the Governance Paradigm* (pp. 24–36). Sydney: Federation Press.

Jinks, P., Biskup, P., & Nelson, H. (1973). *Readings in New Guinea History*. Sydney: Angus and Roberson.

Johnston, M. (1986). Right & Wrong in American Politics: Popular Conceptions of Corruption. *Polity*, 18(3), 367–391.

Johnston, M., & Kpundeh, S. J. (2004). *Building a Clean Machine: Anti-Corruption Coalitions and Sustainable Reform*. World Bank Policy Research Working Paper 3466. Washington, DC: World Bank.

Joku, H. (2010). Icon of Melanesia. 10 March. *PNG Post-Courier*, p. 3.

Jones, A. (2014). *Corruption amongst Government Ministers in Papua New Guinea: Leadership Tribunals and Their Impact on Re-Election*. International IDEA (unpublished).

Justice Advisory Group. (2005). *Fighting Corruption and Promoting Integrity in Public Life in Papua New Guinea*. 20 February. Port Moresby: Justice Advisory Group.

Kanekane, J. (2007). Tolerance and Corruption in Contemporary Papua New Guinea. In A. Ayius & R. J. May (Eds.), *Corruption in Papua New Guinea: Towards an Understanding of Issues* (pp. 23–26). Port Moresby: The National Research Institute.

Keaeke, R. (2003). Papua New Guinea Media Declare War on Corruption. In *Global Corruption Report 2003: The Pacific*. Retrieved 17 January, 2011, from http://unpa n1.un.org/intradoc/groups/public/documents/APCITY/UNPAN008443.pdf, p. 120.

Keane, B. (2010). Who Profits from Our Foreign Aid?: The 'Technical Assistance' Making Business Rich. 12 July. *Crikey*. Retrieved 13 September, 2011, from http:// www.crikey.com.au/2010/07/12/who-profits-from-our-foreign-aid-the-technical-assistance-making-business-rich/

Kelola, T. (2010). Firearm Not to Solve Problem. 23 September. *PNG Post-Courier*, p. 15.

Ketan, J. (2004). *The Name Must Not Go Down: Political Competition and State-Society Relations in Mount Hagen, Papua New Guinea*. Suva: Institute of Pacific Studies, University of the South Pacific.

Ketan, J. (2007). *The Use and Abuse of Electoral Development Funds and Their Impact on Electoral Politics and Governance in Papua New Guinea*. CDI Policy Papers on Political Governance 2007/02. Port Moresby: Centre for Democratic Institutions.

Klitgaard, R. E. (1991). *Controlling Corruption*. Berkeley: University of California Press.

Koim, S. (2013). Turning the Tide: Corruption and Money Laundering in PNG. *Griffith Journal of Law & Human Dignity*, 1(2), 240–253.

Koim, S. (2015). *Progressive Report*. 13 November. Port Moresby: Investigation Taskforce Sweep.

Koim, S., Walton, G. W., & Betteridge, A. (2014). The Perilous State of Taskforce Sweep: An Interview with Sam Koim. *Devpolicy Blog*, 19 December. Retrieved 7 May, 2015, from http://devpolicy.org/the-perilous-state-of-taskforce-sweep-an-interview-with-sam-koim-20141219-2/

Kolao, N. A. (2008). *NGOs and Civil Society Coalition Partners Desk*. Port Moresby: NGOs and Civil Society Coalition.

Krastev, I. (2004). *Shifting Obsessions: Three Essays on the Politics of Anticorruption*. Budapest: CEU Press.

Kurer, O. (2007). Why Do Papua New Guinean Voters Opt for Clientelism? Democracy and Governance in a Fragile State. *Pacific Economic Bulletin*, 22(1), 39–53.

Lambsdorff, J. G. (2007). *The Institutional Economics of Corruption and Reform: Theory, Evidence and Policy*. New York: Cambridge University Press.

Lapauve, J. (2015). Woodlawn Capital Pays K41 Million to MVIL. Retrieved 20 April, 2017, from http://www.emtv.com.pg/news/2015/02/woodlawn-capital-pays-k41-million-to-mvil-2/

Larmour, P. (1997). *Corruption and Governance in the South Pacific*. State, Society and Governance in Melanesia 5. Canberra: Australian National University.

Larmour, P. (2003). Transparency International and Policy Transfer in Papua New Guinea. *Pacific Economic Bulletin*, 18(1), 115–120.

Larmour, P. (2005). *Civilizing Techniques: Transparency International and the Spread of Anti-Corruption*. Asia Pacific School of Economics and Government: Discussion Paper 05. Canberra: Australian National University.

Larmour, P. (2006). *Culture and Corruption in the Pacific Islands: Some Conceptual Issues and Findings from Studies of National Integrity Systems*. Asian Pacific School of Economics and Government: Discussion Paper 06–05. Canberra: Australian National University.

Larmour, P. (2007). International Action against Corruption in the Pacific Islands: Policy Transfer, Coercion and Effectiveness. *Asian Journal of Political Science*, 15(1), 1–16.

Larmour, P. (2012). *Interpreting Corruption: Culture and Politics in the Pacific Islands*. Honolulu: University of Hawaii Press.

Laurance, W. F., Kakul, T., Keenan, R. J., Sayer, J., Passingan, S., Clements, G. R., Villegas, F., & Sodhi, N. S. (2011). Predatory Corporations, Failing Governance, and the Fate of Forests in Papua New Guinea. *Conservation Letters*, 4(2), 95–100.

Leff, N. H. (1964). Economic Development through Bureaucratic Corruption. *The American Behavioral Scientist*, 8(3), 8–14.

Levin, M., & Satarov, G. (2000). Corruption and Institutions in Russia. *European Journal of Political Economy*, 16, 113–132.

McCusker, R. (2006). *Review of Anti-Corruption Strategies*. Technical and Background Paper. Canberra: Australian Government, Australian Institute of Criminology.

McKenzie, N., Baker, R., & Garnaut, J. (2015). Steering Corrupt Cash into Australia from PNG: A How-to Guide. 24 June. *Sydney Morning Herald*. Retrieved 26 June, 2015, from http://www.smh.com.au/national/steering-corrupt-cash-into-australia-from-png-a-howto-guide-20150623-ghv1sx.html

McLeod, A. (2002). Where Are the Women in Simbu Politics? *Development Bulletin*, 59(October), 43–46.

McLeod, S. (2003). Report Outlines Corruption in PNG Immigration System. Retrieved 23 November, 2015, from http://www.abc.net.au/worldtoday/content/2003/s988375.htm

McPhedran, I. (2006). Corrupt Payment for Moti Escape. 17 October. *The Courier-Mail*, p. 12.

Mana, B. (1999). *An Anti-Corruption Strategy for Provincial Government in Papua New Guinea*. Asia Pacific School of Economics and Government: Working Papers, GOV99–5, 1–10.

Marshall, I. E. (2001). *A Survey of Corruption Issues in the Mining and Mineral Sector*. London, UK: Mining, Minerals and Sustainable Development Project.

Mauss, M. (1990). *The Gift: The Form and Reason for Exchange in Archaic Societies*. London: Routledge.

May, R. J. (2004). *State and Society in Papua New Guinea: The First Twenty-Five Years* (2nded.). Canberra: Australian National University E Press.

May, R. J. (2006). The 'Clan Vote' in Papua New Guinea: Data from Angoram. *The Journal of Pacific Studies*, 29(1), 108–129.

Mbaku, J. M. (2007). *Corruption in Africa: Causes, Consequences and Cleanups*. Plymouth: Lexington Books.

Mellam, A., & Aloi, D. (2003). *Country Study Report: Papua New Guinea 2003*. National Integrity Systems. Berlin: Transparency International.

Merrell, S. (2012). Democracy, Custom & the Elasticity of the Melanesian Way. 21 August. Retrieved 22 July, 2015, from http://asopa.typepad.com/asopa_people/2012/08/democracy-custom-the-elasticity-of-the-melanesian-way.html

Michael, B. (2004). What Do African Donor-Sponsored Anti-Corruption Programmes Teach Us about International Development in Africa? *Social Policy and Administration*, 38(4), 320–345.

Michael, B., & Bowser, D. (2009). The Evolution of the Anti-Corruption Industry in the Third Wave of Anti-Corruption Work. Paper presented at the Konstanz Anti-Corruption Conference, Konstanz.

Millennium Good Governance. (2006). *Petition against Corruption*. (31st Independence Anniversary). Port Moresby: Millennium Good Governance.

Minta, P. (2014). *The Prevalence of Transnational Crimes in Papua New Guinea: By Design or by Default?* State, Society & Governance in Melanesia Program. Canberra: Australian National University.

Moroff, H., & Schmidt-Pfister, D. (2010). Anti-Corruption Movements, Mechanisms, and Machines – an Introduction. *Global Crime*, 11(2), 89–98.

Mungiu-Pippidi, A. (2013). Controlling Corruption through Collective Action. *Journal of Democracy*, 24(1), 101–115.

Murphy, J. (2011). Capitalism and Transparency. *Critical Perspectives on International Business*, 7(2), 125–141.

Myrdal, G. (1968). *Asian Drama: An Inquiry into the Poverty of Nations*. New York: Pantheon Books.

Namorong, M. (2012). Transparency Drags Feet over Democracy March. 9 April. Retrieved 23 November, 2016, from http://asopa.typepad.com/asopa_people/2012/04/transparency-drags-feet-over-png-democracy-march-which-pm-oneill-calls-illegal.html

Narokobi, B. (1980). *The Melanesian Way*. Boroko, Papua New Guinea: Institute of Papua New Guinea Studies.

Narokobi, B. (1989). *Lo Bilong Yumi Yet: Law and Custom in Melanesia*. Goroka: Melanesian Institute for Pastoral and Socio-Economic Service and The University of the South Pacific.

NASFUND. (n.d.). About Nasfund. Retrieved 4 February, 2011, from http://www.nasfund.com.pg/index.php/about/vision-mission

National Provident Fund Final Report. (2015). National Provident Fund Final Report [Part 69]. Retrieved 9 December, 2016, from https://pngexposed.wordpress.com/2015/11/10/national-provident-fund-final-report-part-69/

Nederveen Pieterse, J. (2000). After Post-Development. *Third World Quarterly*, 21(2), 175–191.

Nelson, H. (1974). *Papua New Guinea: Black Unity or Black Chaos?* Melbourne: Penguin Books.

Noonan, J. T. (1984). *Bribes*. New York: Macmillan.

Nuijten, M., & Anders, G. (2007). Corruption and the Secret of Law: An Introduction. In M. Nuijten & G. Anders (Eds.), *Corruption and the Secret of Law: A Legal Anthropological Perspective* (pp. 1–26). Aldershot, UK: Ashgate.

Nye, J. S. (1967). Corruption and Political Development: A Cost-Benefit Analysis. *American Political Science Review*, 61(2), 417–427.

Oakland Institute. (2015). *The Great Timber Heist: The Logging Industry in Papua New Guinea*. Oakland, CA: The Oakland Institute.

OECD Working Group on Bribery. (2012). *Phase 3 Report on Implementing the OECD Anti-Bribery Convention in Australia*. October. OECD.

Office of Development Effectiveness. (2007). *Approaches to Anti-Corruption through the Australian Aid Program: Lessons from PNG, Indonesia and the Solomon Islands*. Canberra: Australian Government.

Office of the Auditor-General of Papua New Guinea. (2010). *Part II – Report of the Auditor-General – 2008 National Government Departments and Agencies: On the Controls in 2009 and on Transactions with or Concerning the Public Monies and Property of PNG in 2008*. Port Moresby: Auditor-General's Office Papua New Guinea.

Ohnesorge, J. K. M. (1999). Ratcheting up the Anti-Corruption Drive: Could a Look at Recent History Cure a Case of Theory-Determinism? *Connecticut Journal of International Law*, 14, 467–473.

Okole, H., & Kavanamur, D. (2003). Political Corruption in Papua New Guinea: Some Causes and Policy Lessons. *South Pacific Journal of Philosophy and Culture*, 7(1), 7–36.

Ombudsman Commission of Papua New Guinea. (1999). *Investigation into the Purchase of the Conservatory, Cairns by the Public Officers Superannuation Fund Board*. Port Moresby: Ombudsman Commission of Papua New Guinea.

Ombudsman Commission of Papua New Guinea. (2001). *Investigations Practice Manual*. Port Moresby: Ombudsman Commission of Papua New Guinea.

Ombudsman Commission of Papua New Guinea. (2006). *Annual Report 2005: For the Period of 1 January 2005 to 31 December 2005*. Port Moresby: Ombudsman Commission of Papua New Guinea.

Organisation for Economic Co-operation and Development. (2010). *The State's Legitimacy in Fragile Situations: Unpacking Complexity*. Conflict and Fragility. OECD.

Oxford Dictionaries. (n.d.). Corruption. Retrieved 30 January, 2012, from http://oxforddictionaries.com/definition/english/corruption?q=corruption

Palme, R. (2008). Somare U-Turn on Fin Inquiry. 23 April. *PNG Post-Courier*, p. 1.

Parkinson, J. (2003). Legitimacy Problems in Deliberative Democracy. *Political Studies*, 51(1), 180–196.

Payani, H. H. (2000). Selected Problems in the Papua New Guinean Public Service. *Asian Journal of Public Administration*, 22(2), 135–160.

Pelto, M. (2007). Civil Society and the National Integrity System in Papua New Guinea. *Pacific Economic Bulletin*, 22(1), 54–69.

Perry, P. J. (1997). *Political Corruption and Political Geography*. Aldershot: Ashgate.

Perry, P. J. (2001). *Political Corruption in Australia: A Very Wicked Place?* Aldershot: Ashgate.

Peters, J., & Welch, S. (1978). Political Corruption in America: A Search for Definitions and a Theory, or If Political Corruption Is in the Mainstream of American Politics Why Is It Not in the Mainstream of American Politics Research? *The American Political Science Review*, 72(3), 974–984.

Pitts, M. (2001). Crime and Corruption – Does Papua New Guinea Have the Capacity to Control It? *Pacific Economic Bulletin*, 16(2), 127–134.

Pitts, M. (2002). *Crime, Corruption and Capacity in Papua New Guinea*. Canberra: Asia Pacific Press.

PNG Justice Advisory Group. (2007). *Port Moresby Community Crime Survey, 2006*. Port Moresby: PNG Justice Advisory Group.

PNG National Awareness Front, Eda Hanua Moresby Inc., PNG Millennium Good Governance Organisation, & NCD Youths Association. (2007). *Media Press Statement*. Port Moresby: NGOs and Civil Society Desk.

PNG Post-Courier. (1990a). Students Plan Major Protest against the Government. 24 July. *PNG Post-Courier*, p. 3.

PNG Post-Courier. (1990b). More Protests to Fight Corruption. 1 August. *PNG Post-Courier*, p. 10.

PNG Post-Courier. (2003). PNG Corruption 'Simply Not Homegrown'. Retrieved 19 December, 2006, from http://www.postcourier.com.pg/20030429/news09

PNG Post-Courier. (2006). K3m Settlement. 22 May. *PNG Post-Courier*, p. 5.

PNG Post-Courier. (2008). T.I. Blasts Govt Action. 21 April. *PNG Post-Courier*, p. 6.

PNG Post-Courier. (2009). Six Pack for Passport Deals PM. 26 May. *PNG Post-Courier*, p. 6.

PNG Post-Courier. (2012a). Aliens in PNG. 13 January. *PNG Post-Courier*, p. 3.

PNG Post-Courier. (2012b). Thousands Turn Up. 11 April. *PNG Post-Courier*, p. 4.

PNG Post-Courier. (2013a). PNG Opposition Concerned by Chief Ombudsman Vacancy. 12 June. Retrieved 24 November, 2016, from http://pidp.org/pireport/2013/June/06-13-15.htm

PNG Post-Courier. (2013b). TI PNG Bid on Polls. 12 March. *PNG Post-Courier*, p. 6.

PNG Post-Courier. (2015). Tiensten Gets Three More Years in Prison. 17 April. *PNG Post Courier*, p. 3.

Pok, J. (2014). Tiensten Jailed MP Gets Nine Years for K10m Payment; Four Years Will Be Cut If He Repays State. 31 March. *PNG Post-Courier*, p. 1.

Pok, J. (2015). Kondra Faces Tribunal. 20 February. *PNG Post-Courier*, p. 9.

Polinksy, A. M., & Shavell, S. (2001). Corruption and Optimal Law Enforcement. *Journal of Public Economics*, 81(1), 1–24.

Polzer, T. (2001). *Corruption: Deconstructing the World Bank Discourse*. Working Paper No. 01–18. London: Development Studies Institute.

Pope, J. (2000a). *Confronting Corruption: The Elements of a National Integrity System. TI Source Book 2000*. Berlin: Transparency International.

Pope, J. (Ed.). (2000b). *National Integrity Systems: The TI Source Book*. Berlin: Transparency International.

Posadas, A. (2000). Combatting Corruption under International Law. *Duke Journal of Comparative and International Law*, 10(2), 335–414.

Prime Minister's Office. (2013). Prime Ministerial Media Release. Retrieved 16 February, 2016, from http://asopa.typepad.com/asopa_people/2013/03/election-was-fine-transpa rency-exaggerated-says-oneill-gulp-peter-oneill-that-seems-a-bit-rough.html

Qizilbash, M. (2001). Corruption and Human Development: A Conceptual Discussion. *Oxford Development Studies*, 29(3), 265–278.

Radio New Zealand. (2010). PNG Ombudsman to Investigate Activist's Claim He Was Attacked by Somare Family. Retrieved 24 November, 2016, from http://www.radionz. co.nz/international/pacific-news/194488/png-ombudsman-to-investigate-activist's-claim-he-was-attacked-by-somare-family

Radio New Zealand. (2016). Concerns Raised about New PNG Cyber Crime Laws. 12 August. Retrieved 9 December, 2016, from http://www.radionz.co.nz/international/ pacific-news/310837/concerns-raised-about-new-png-cyber-crime-laws

Radio New Zealand International. (2003). Melanesian Solidarity Spokesman Says Ausaid Carries Some Responsibility for PNG's Problems. 19 September. Retrieved 19 September, 2012, from http://www.rnzi.com/pages/news.php?op=read&id=6569

Radio New Zealand International. (2008). PNG Police Imposters to Be Shot on Sight, Says Commissioner. 9 July. Retrieved 12 July, 2010, from http://www.rnzi.com/ pages/news.php?op=read&id=40821

Radio New Zealand International. (2009). PNG Group Opposed to Asian Immigrants but Vows It Is Not Racist. 14 May. Retrieved 24 August, 2015, from http://www.ra dionz.co.nz/international/pacific-news/183694/png-group-opposed-to-asian-immigra nts-but-vows-it-is-not-racist

Radio New Zealand International. (2012). High Profile PNG Civil Society Activist Dumped as Leader. 16 April. Retrieved 24 August, 2015, from http://www.radionz.co.nz/interna tional/pacific-news/203921/high-profile-png-civil-society-activist-dumped-as-leader

Rahnema, M. (1992). Poverty. In W. Sachs (Ed.), *The Development Dictionary: A Guide to Knowledge as Power* (pp. 158–176). London: Zed Books.

Raitano, S. (2015). Knight Found Guilty on Four Misuse Charges. 23 March. *PNG Post-Courier*, p. 5.

Rannells, J., & Matatier, E. (2005). *PNG Fact Book: A One-Volume Encyclopedia of Papua New Guinea* (3rded.). South Melbourne: Oxford University Press.

Reilly, B. (1999). Party Politics in Papua New Guinea: A Deviant Case? *Pacific Affairs*, 72(2), 225–246.

Reilly, B. (2000). The Africanisation of the South Pacific. *Australian Journal of International Affairs*, 54(3), 261–268.

Rheeney, A. (2005). MPs Back Chief's Call. 21 October. *PNG Post-Courier*, p. 8.

Rheeney, A., & Sasere, E. (2002). TI: 'It's Stealing'. 2 September. *PNG Post-Courier*, p. 1.

Richards, K. (2006). *What Works and Why in Community-Based Anti-Corruption Programs*. Melbourne: Transparency International Australia.

Robbins, J. (1998). On Reading 'World News': Apocalyptic Narrative, Negative Nationalism and Transnational Christianity in a Papua New Guinea Society. *Social Analysis*, 42(2), 103–130.

Roberts, G. (2006). The Rape of PNG's Forests. 24 June. *The Australian*. Retrieved 3 February, 2007, from http://www.theaustralian.com.au/story/0,20867,19565361-30417,00.html

Rocha, L. J., Brown, E., & Cloke, J. (2011). Of Legitimate and Illegitimate Corruption: Bankruptcies in Nicaragua. *Critical Perspectives on International Business*, 7(2), 159–176.

Rose-Ackerman, S. (1999). *Corruption and Government: Causes, Consequences and Reform*. New York: Cambridge University Press.

Ruud, A. E. (2000). Corruption as Everyday Practice. The Public-Private Divide in Local Indian Society. *Forum for Development Studies*, 27(2), 271–294.

Sachs, W. (1992). Introduction. In W. Sachs (Ed.), *The Development Dictionary: A Guide to Knowledge as Power* (pp. 1–5). Johannesburg: Witwatersrand University Press, Zed Books.

Sahlins, M. (1972). *Stone Age Economics*. London: Routledge.

Sampson, S. (2005). Integrity Warriors: Global Morality and the Anti-Corruption Movement in the Balkans. In D. Haller & C. Shore (Eds.), *Corruption: Anthropological Perspectives* (pp. 103–130). London: Pluto.

Sampson, S. (2010). The Anti-Corruption Industry: From Movement to Institution. *Global Crime*, 11(2), 261–278.

Scott, J. C. (1972). *Comparative Political Corruption*. Englewood Cliffs, NJ: Prentice-Hall.

Sen, A. (1999). *Development as Freedom*. Oxford: Oxford University Press.

Sepoe, O. (2002). To Make a Difference: Realities of Women's Participation in PNG Politics. *Development Bulletin*, 59, 39–42.

Sharman, J. C. (2012). Chasing Kleptocrats' Loot: Narrowing the Effectiveness Gap. Anti-Corruption Resource Centre. Retrieved 13 January 2015, from http://www.u4. no/publications/chasing-kleptocrats-loot-narrowing-the-effectiveness-gap/

Sharp, A. (2016). The Challenges of Australia's Northern Border: The Torres Strait. *The Strategist*, 10 June. Retrieved 17 October, 2016, from http://www.aspistrategist. org.au/challenges-australias-northern-border-torres-strait/

Shore, C., & Wright, S. (1997). Policy: A New Field of Anthropology. In C. Shore & S. Wright (Eds.), *Anthropology of Policy: Critical Perspectives on Governance and Power* (pp. 3–42). London and New York: Routledge.

Skinner, Q. (1990). The Republican Ideal of Political Liberty. In G. Bock (Ed.), *Machiavelli and Republicanism* (pp. 293–309). Cambridge: Cambridge University Press.

Smith, G. (2012). Chinese Reactions to Anti-Asian Riots in the Pacific. *The Journal of Pacific History*, 47(1), 93–109.

Smith, S., & McMullan, B. (2009). *Budget: Australia's International Development Assistance Program, a Good International Citizen*. Canberra: Commonwealth of Australia.

Sööt, M.-L., & Rootalu, K. (2012). Institutional Trust and Opinions of Corruption. *Public Administration and Development*, 32(1), 82–95.

South Pacific Post. (2010). "Prime Minister Ordered My Kidnapping" Noel Anjo. Retrieved 23 November, 2016, from http://www.pngblogs.com/2010/11/prime-minis ter-ordered-my-kidnapping.html

Standish, B. (1999). *Papua New Guinea 1999: Crisis of Governance*. Research Paper 4 1999–2000. 21 September. Canberra: Foreign Affairs, Defence and Trade Group.

Standish, B. (2007). The Dynamics of Papua New Guinea's Democracy: An Essay. *Pacific Economic Bulletin*, 22(1), 135–156.

Stewart, P. J., & Strathern, A. J. (1998). Money, Politics, and Persons in Papua New Guinea. *Social Analysis*, 42(2), 132–149.

Syndicus, I. (2015). Student Boycotts at the University of Goroka: Some Patterns and Distinct Features. *Devpolicy Blog*, 9 September. Retrieved 24 November, 2016, from http://devpolicy.org/student-boycotts-at-the-university-of-goroka-some-patterns-and-distinct-features-20150909/

Szeftel, M. (1998). Misunderstanding African Politics: Corruption and the Governance Agenda. *Review of African Political Economy*, 25(76), 221–240.

Temby, S. (2007). *Good Governance in Papua New Guinea: An Australian Agenda.* School of Social and Environmental Enquiry Research Papers. Melbourne: The University of Melbourne.

The Asia-Pacific Action Group. (1990). *The Barnett Report: A Summary of the Report of the Commission of Inquiry into Aspects of the Timber Industry in Papua New Guinea.* Hobart: The Asia-Pacific Action Group.

The Government of Australia & The Government of Papua New Guinea. (2008). *Partnership for Development between the Government of Australia and the Government of Papua New Guinea.* Retrieved 20 April, 2017, from http://webarchive.nla. gov.au/gov/20120322063412/http://www.ausaid.gov.au/publications/pubout.cfm?ID= 3419_4132_845_3563_3801&Type=

The National. (2009). History Critical for PNG Nation-Building AusAid Told. 18 May. Retrieved 13 November, 2014, from www.thenational.com.pg

Theobald, R. (1990). *Corruption, Development and Underdevelopment.* Basingstoke: Macmillan.

Thomas, M. A., & Meagher, P. (2004). *A Corruption Primer: An Overview of Concepts in the Corruption Literature.* February. College Park, MD: IRIS Center.

Transparency International. (2009). *Global Corruption Barometer.* Berlin: Transparency International.

Transparency International. (2015). *Corruption Perceptions Index 2015.* Berlin: Transparency International, International Secretariat.

Transparency International. (n.d.-a). Frequently Asked Questions About Corruption. Retrieved 15 December, 2014, from http://www.transparency.org/whoweare/organisa tion/faqs_on_corruption/2/

Transparency International. (n.d.-b). Our Organisation. Retrieved 20 April, 2017, from http://www.transparency.org/whoweare/organisation/

Transparency International Papua New Guinea. (2005). *Join the Fight against Corruption.* Port Moresby: Transparency International Papua New Guinea.

Transparency International Papua New Guinea (Ed.). (2007). *10 Years of Fighting Corruption 1997–2007.* Port Moresby: Transparency International Papua New Guinea.

Transparency International Papua New Guinea. (2008a). *Annual Report 2007.* Port Moresby: Transparency International Papua New Guinea.

Transparency International Papua New Guinea. (2008b). *K1 Billion Missing from Government Coffers.* Press Release. Port Moresby: Transparency International Papua New Guinea.

Transparency International Papua New Guinea. (2008c). *Procurement and Financial Management.* Press Release. Port Moresby: Transparency International Papua New Guinea.

Transparency International Papua New Guinea. (2008d). *Walk against Corruption.* Port Moresby: Transparency International Papua New Guinea.

Transparency International Papua New Guinea. (2009a). About Us. Retrieved 24 November, 2016, from http://www.transparencypng.org.pg/about_us.html

Transparency International Papua New Guinea. (2009b). *Annual Report 2008.* Port Moresby: Transparency International Papua New Guinea.

Transparency International Papua New Guinea. (2012a). TI Papua New Guinea: Judiciary Conduct Act 2012 Must Be Repealed Immediately. 27 March. Retrieved 24 November, 2016, from http://www.transparency.org/news/pressrelease/20120327_ TI_papue_new_guinea

Transparency International Papua New Guinea. (2012b). *TI PNG Observation Report: The 2012 National Parliamentary Elections.* Port Moresby: Transparency International Papua New Guinea.

Treisman, D. (2000). The Causes of Corruption: A Cross-National Study. *Journal of Public Economics*, 76(3), 399–457.

True, J., Niner, S., Parashar, S., & George, N. (2012). *Women's Political Participation in Asia and the Pacific.* New York: Social Science Research Council's Conflict Prevention & Peace Forum, and the United Nations Department of Political Affairs.

Ugur, M., & Dasgupta, N. (2011). *Evidence on the Economic Growth Impacts of Corruption in Low-Income Countries and Beyond: A Systematic Review.* London: EPPI-Centre, Social Science Research Unit, Institute of Education, University of London.

United Nations Development Program. (2009). PNG Businesses Fight Corruption. 27 August. Port Moresby: United Nations Development Program.

United Nations Development Program. (2015). *Human Development Report 2015: Work for Human Development.* New York: United Nations Development Program.

UNODC. (2013). *Transnational Organized Crime in East Asia and the Pacific.* Bangkok: UNODC.

Van Rijckeghem, C., & Weder, B. (2001). Bureaucratic Corruption and the Rate of Temptation: Do Wages in Civil Service Affect Corruption, and by How Much? *Journal of Development Economics*, 65(2), 307–331.

Vorkink, A. (2000). Corruption, a Brake to Development! Paper presented at the Regional Conference on Anti-Corruption, Bucharest, 30 March. World Bank.

Walton, G. W. (2012). *Comparing Local and International Perspectives on Corruption in Papua New Guinea.* PhD thesis. Department of Resource Management and Geography, The University of Melbourne, Melbourne.

Walton, G. W. (2013a). Anti-Corruption on the Front Line: An Interview with Sam Koim. *Devpolicy Blog*, 13 June. Retrieved 25 June, 2015, from http://devpolicy.org/anti-corruption-on-the-front-line-an-interview-with-sam-koim-20130611/

Walton, G. W. (2013b). Is All Corruption Dysfunctional? Perceptions of Corruption and Its Consequences in Papua New Guinea. *Public Administration and Development*, 33 (3), 175–190.

Walton, G.W. (2013c). An Argument for Reframing Debates About Corruption: Insights from Papua New Guinea. *Asia Pacific Viewpoint*, 54(1), 61–76.

Walton, G. W. (2015). Defining Corruption Where the State Is Weak: The Case of Papua New Guinea. *Journal of Development Studies*, 51(1), 15–31.

Walton, G. W. (2016). Gramsci's Activists: How Local Civil Society Is Shaped by the Anti-Corruption Industry, Political Society and Trans-local Encounters. *Political Geography*, 53, 10–19.

Walton, G. W., & Dinnen, S. (2016). *The Dark Side of Economic Globalisation: Politics, Organised Crime and Corruption in the Pacific.* Development Policy Centre Discussion Paper No. 48. Canberra: Australian National University.

Walton, G. W., & Howes, S. (2014). Using the C-Word: Australian Anti-Corruption Policy in Papua New Guinea. *Devpolicy Blog*, 22 August. Retrieved 2 September, 2015, from http://devpolicy.org/using-the-c-word-australian-anti-corruption-policy-in-papua-new-guinea-20140822/

Walton, G. W., & Peiffer, C. (2015). *The Limitations of Education for Addressing Corruption: Lessons from Attitudes towards Reporting in Papua New Guinea.* Crawford School Development Policy Centre Discussion Paper 39. 1 June. Retrieved 4 September, 2015, from http://ssrn.com/abstract=2614179

Warr, P. (1997). No Clear Objective: The Simons Report on Foreign Aid. *Agenda*, 5(3), 362–366.

Webster, C. (2010). PNG Inquiry Reveals Finance Department Corruption. *Pacific Beat*. 5 March. Retrieved 12 May, 2010, from http://www.radioaustralia.net.au/pacbeat/stories/201003/s2838082.htm

Wenambo, N. (2015). Education Dept Signs Deal. 15 June. *PNG Post-Courier*, p. 12.

Wilkinson, M., & Cubby, B. (2009). Carbon Scandal Snares Australian. 4 September. *Sydney Morning Herald*. Retrieved 21 November, 2015, from http://www.smh.com.au/business/carbon-scandal-snares-australian-20090903-f9yy.html

Williams, R. (1999). New Concepts for Old? *Third World Quarterly*, 20(3), 503–513.

Windybank, S., & Manning, M. (2003). *Papua New Guinea on the Brink*. Issue Analysis 30. St Leonards: The Centre for Independent Studies.

Wood, J. T. (n.d.). *Leadership Codes and Corruption Prevention*. Retrieved 13 August, 2015, from https://dfat.gov.au/about-us/corporate/freedom-of-information/Documents/leadership-codes-corruption-prevention.pdf

World Bank. (1997). *Helping Countries Combat Corruption: The Role of the World Bank*. Washington, DC: World Bank.

World Bank. (2001). World Bank, IMF Escalate War against Corruption. *Beyond Transition: The Newsletter About Reforming Economies*. Retrieved 2 December, 2010, from http://wwwuser.gwdg.de/~uwvw/oldwebsite/Press97/worldbnk.htm

World Bank. (2007). *Strengthening World Bank Group Engagement on Governance and Anti-Corruption*. Washington, DC: World Bank.

World Bank. (2010). Six Questions on the Cost of Corruption with World Bank Institute Global Governance Director Daniel Kaufmann. *News and Broadcast*. Retrieved 23 November, 2010, from http://web.worldbank.org/WBSITE/EXTERNAL/NEWS/0,,contentMDK:20190295~menuPK:34457~pagePK:34370~piPK:34424~theSitePK:4607,00.html

World Bank. (2017). World Data Bank: Papua New Guinea. Retrieved 4 June, 2017, from http://databank.worldbank.org/data

Yang, M. (1989). The Gift Economy and State Power in China. *Comparative Studies in Society and History*, 31(1), 25–54.

Yang, M. (1994). *Gifts, Favours and Banquets: The Art of Social Relationships in China*. London: Cornell University Press.

Index

The acronym PNG is used throughout for Papua New Guinea. Locators in **bold** refer to tables and those in *italics* to figures.